HANDBOOKS

❖

NUMBER NINE

Topics in Combinatorics

Gerry Leversha
and
Dominic Rowland

United Kingdom
Mathematics Trust

Topics in Combinatorics

Published by the United Kingdom Mathematics Trust.

School of Mathematics, University of Leeds,

Leeds, LS2 9JT, United Kingdom

www.ukmt.org.uk

First published 2019.

ISBN 978-1-906001-36-0

Printed in the UK for the UKMT by Charlesworth Press, Wakefield.

www.charlesworth.com

Typographic design by Andrew Jobbings of Arbelos.

www.arbelos.co.uk

Typeset with LaTeX.

The books published by the United Kingdom Mathematics Trust are grouped into series.

The EXCURSIONS IN MATHEMATICS series consists of monographs which focus on a particular topic of interest and investigate it in some detail, using a wide range of ideas and techniques. They are aimed at high school students, undergraduates and others who are prepared to pursue a subject in some depth, but do not require specialised knowledge.

1. *The Backbone of Pascal's Triangle*, Martin Griffiths
2. *A Prime Puzzle*, Martin Griffiths

The HANDBOOKS series is aimed particularly at students at secondary school who are interested in acquiring the knowledge and skills which are useful for tackling challenging problems, such as those posed in the competitions administered by the UKMT and similar organisations.

1. *Plane Euclidean Geometry: Theory and Problems*, A D Gardiner and C J Bradley
2. *Introduction to Inequalities*, C J Bradley
3. *A Mathematical Olympiad Primer*, Geoff C Smith
4. *Introduction to Number Theory*, C J Bradley
5. *A Problem Solver's Handbook*, Andrew Jobbings
6. *Introduction to Combinatorics*, Gerry Leversha and Dominic Rowland
7. *First Steps for Problem Solvers*, Mary Teresa Fyfe and Andrew Jobbings
8. *A Mathematical Olympiad Companion*, Geoff C Smith
9. *Topics in Combinatorics*, Gerry Leversha and Dominic Rowland

The PATHWAYS series aims to provide classroom teaching material for use in secondary schools. Each title develops a subject in more depth and in more detail than is normally required by public examinations or national curricula.

1. *Crossing the Bridge*, Gerry Leversha
2. *The Geometry of the Triangle*, Gerry Leversha

The PROBLEMS series consists of collections of high-quality and original problems of Olympiad standard.

1. *New Problems in Euclidean Geometry*, David Monk

The CHALLENGES series is aimed at students at secondary school who are interested in tackling stimulating problems, such as those posed in the Mathematical Challenges administered by the UKMT and similar organisations.

1. *Ten Years of Mathematical Challenges: 1997 to 2006,*
2. *Ten Further Years of Mathematical Challenges: 2006 to 2016,*
3. *Intermediate Problems*, Andrew Jobbings
4. *Junior Problems*, Andrew Jobbings
5. *Senior Problems*, Andrew Jobbings

The YEARBOOKS series documents all the UKMT activities, including details of all the challenge papers and solutions, lists of high scorers, accounts of the IMO and Olympiad training camps, and other information about the Trust's work during each year.

Contents

Series Editor's Foreword

This book is part of a series whose aim is to help young mathematicians prepare for competitions, such as the British Mathematical Olympiad, at secondary school level. Like other volumes in the Handbooks series, it provides cheap and ready access to directly relevant material. All these books are characterised by the large number of carefully constructed exercises for the reader to attempt.

I hope that every secondary school will have these books in its library. The prices have been set so low that many good students will wish to purchase their own copies. Schools wishing to give out large numbers of copies of these books, perhaps as prizes, should note that discounts may be negotiated with the UKMT office.

London, UK GERRY LEVERSHA

About the Authors

Gerry Leversha taught mathematics in secondary schools before retiring in 2011. He has also been involved in the work of the UKMT, both in the setting and marking of various Olympiads and as Chair of the Publications Committee. He is also the editor of *The Mathematical Gazette*, the journal of the Mathematical Association, and is a regular speaker at conferences and summer schools. His interests include music, film and literature, wine and cooking as well as playing tennis and mountain walking.

Dominic Rowland participated in his first UKMT maths challenge when he was eight years old. Over the next decade he participated in various challenges, Olympiads and maths camps. He began teaching mathematics at secondary school in 2011 and is a regular speaker at UKMT summer schools and mathematical circles as well as helping to set and mark the British Mathematical Olympiads. He enjoys board games, walking and cycling.

Preface

In the preface to *Introduction to Combinatorics* ([4]) we expressed a hope to produce a follow-up volume, and this book is the realisation of that hope. As promised, we are now able to discuss graph theory, Ramsey theory, Pólya counting and much more besides.

Like its predecessor, this book is problem-based and contains a large number of exercises for you to try. We firmly believe that mathematics is not a spectator sport, and we hope you will enjoy working on the questions as you read.

As authors, our aim is not only to explain what the solutions to the various problems are, but also *how these solutions might be found*. Sometimes this has meant discussing plausible wrong turns on the path to a solution, or highlighting the role of reasoning by analogy and following rules of thumb. Our hope is that this more discursive style (which extends to the solutions to the exercises) will give a sense that even great leaps of insight are often composed of numerous small and natural steps.

A key difference between this book and [4] is that that our choice of material is not guided by the questions often found in Mathematical Olympiads, and we have therefore been able to follow our own interests. The basics of graph theory could not reasonably have been omitted, but the rest of the book forms a smörgåsbord of largely independent chapters. Not every reader will agree with every decision we have made about what to include, but we hope that on the whole the selection will be to your taste.

Our target audience includes ambitious secondary school students and their teachers, though much of the material would complement rather than overlap with an undergraduate pure mathematics course. We have aimed to make this volume fairly self-contained: some familiarity with basic counting techniques, the pigeonhole principle and proof by induction is assumed, and all other notation and theory is introduced in the main text or appendix. Certainly readers who have some familiarity with [4] know more than enough to get started.

Note on historical context

The first two problems in this book concern disjoint sets X and Y and a number of pairs (x_i, y_i) where $x_i \in X$ and $y_i \in Y$. Our first problem was first posed in the 19th century by Édouard Lucas who thought of X as a set of men, Y as a set of women and each pair (x_i, y_i) as a married couple. We have decided to retain this historical phrasing. Our intention in discussing *men* and *women* rather than members of abstract sets is merely to help in the exposition of the *mathematics*. This remark also applies to our second problem, where we have retained the 19th century context together with the definition of *couple* common at that time.

Acknowledgements

This book would have been impossible without the support of a large number of people. We would like to thank all those who read portions of the manuscript providing valuable feedback and helping to reduce the number of errors, particularly: Lennie Wells, Tom Foster, Emily Beatty and Reuel Armstrong. We would also like to thank Daniel Kirk for his help in preparing the typescript, Claire Hall for her punctilious proofreading and Andrew Jobbings for his work on the layout and formatting of this book and its prequel.

London, UK GERRY LEVERSHA

In addition to the above I would like to thank my wife Camilla for her unfailing support. Thanks also to James for your encouragement and to Henry for interrupting the project by your birth.

Winchester, UK DOMINIC ROWLAND

Chapter 1

Social Etiquette

Readers who wish to expand their knowledge of the branch of mathematics known as combinatorics might be surprised at the title of this first chapter. The authors are aware, however, that young people whose major obsession is mathematics are sometimes criticised for their lack of polish. We begin, therefore, by focusing on a few essential skills of etiquette: hosting a dinner party, choosing your partner for life and walking to church in a civilised manner.

1.1 Table manners

As every society hostess knows, the organisation of a successful dinner party involves some essential principles. These include arranging her guests around the table in such a way that the conversation is stimulating and inclusive. To facilitate this, the use of a round table is recommended. Having equal numbers of men and women is a good idea, and one way of achieving this aim, as well as excluding the riff-raff, is to invite only respectable married couples, placing them with the sexes alternating around the table. Finally, it is clearly a good idea to separate husbands and wives so as to minimise the danger of domestic argument.

These desirable aspirations motivate our first problem.

Problem 1.1

In how many ways can six married couples be arranged around a circular table, so that men and women alternate and nobody sits next to their spouse?

In the world of combinatorics, couples consist of one man and one woman, and round tables are assumed to be floating around in a cosmic void, so that no particular seat is special. The number of ways of seating twelve people along a bench is 12!, since there is someone on the left-hand end, but the number of ways of seating them around a round table with no distinguished seat is 11!.

There are two extra conditions in the problem: the guests are seated with men and women alternating and married couples do not sit in adjacent chairs. At this point you should be warned that the calculation which follows is not straightforward; prepare to be patient. It is also a good idea for you to build up the diagrams showing the arrangements around the table step by step, so that you understand the rationale for the assertions which are made.

It is sensible to work up to six couples by considering smaller numbers, but, at the same time, employ methods which can easily be generalised to n couples. Denote the wives by $W_1, W_2, \cdots, W_n$ and their respective husbands by $M_1, M_2, \cdots, M_n$, and call the number of arrangements which satisfy all the conditions of the problem $m(n)$.

It might seem that a good way to proceed is to sit the ladies first (remember your etiquette!) and then fit the gentlemen in. As usual with circular arrangements, we can place W_1 first and then arrange the other wives anticlockwise around the table; this can clearly be done in $(n-1)!$ ways. Now we have n spaces to fill with men, ensuring that no husband is next to his wife. If we denote the number of ways of doing this by $m^*(n)$, then it is clear that $m(n) = (n-1)! \times m^*(n)$.

Another useful observation is that, whatever the arrangement of the wives, we can simply go round the table from W_1 and renumber the successive wives as $W_2, W_3, \cdots, W_n$ (and their husbands accordingly). Therefore $m^*(n)$ is the number of ways of fitting the symbols M_1 to M_n into the spaces in this circular arrangement so that no neighbouring symbols have the same subscript.

A moment's thought should convince you that $m^*(1) = m^*(2) = 0$, since in neither case is it possible to put the men anywhere.

When there are three couples, there is only one way to place the men. In figure 1.1 the spaces between the ladies are labelled using Roman numerals. M_1 must sit in seat II, M_2 in seat III and M_3 in seat I, so there is only one arrangement, with husbands and wives sitting opposite one another. So $m^*(3) = 1$ and $m(3) = 2! \times 1 = 2$.

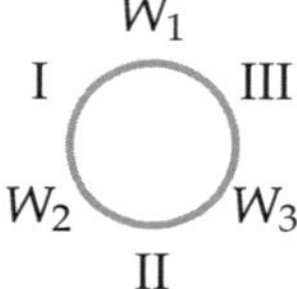

Figure 1.1

Next we consider four couples. We seat the ladies, and label the seats next to them as shown in figure 1.2.

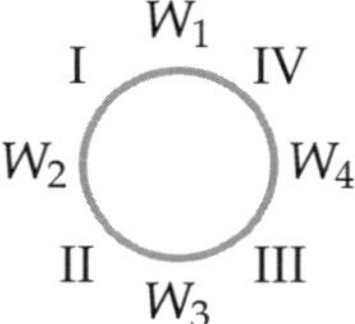

Figure 1.2

There are two cases based on which seat M_1 occupies. If he takes seat II, then M_4 must take seat I, and there is one way to complete the arrangement. If, on the other hand, M_1 takes seat III, then M_2 must take seat IV, which gives rise to one other arrangement.

This seems to be going swimmingly, since the position of M_1 fixes the positions for the other gentlemen. However, when there are five couples, as shown in figure 1.3, this begins to go wrong.

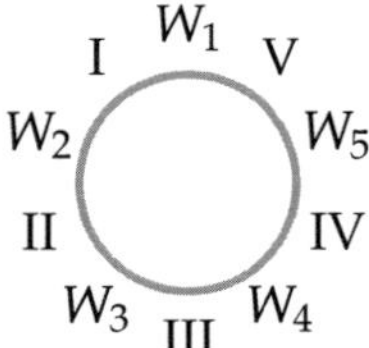

Figure 1.3

To start with there are three cases based on where M_1 sits. However, if M_1 takes seat II, things get a little more complicated. There are three subcases depending on where M_2 sits. If M_1 takes seat III, there are two subcases based on where M_2 sits and these lead to a total of five arrangements. Finally if M_1 takes seat IV, there are again three subcases, which give a total of four arrangements.

At the end of all this, we see that $m^*(5) = 4 + 5 + 4 = 13$ and so $m(5) = 4! \times 13 = 312$.

The trouble with this approach is that we seem to be sliding into *ad hoc* arguments. As n increases, these are going to get more and more complicated, and we will have to consider more and more special cases. Let us go back to the drawing board and start again.

Rather than counting by *addition* — considering what choices there are as men are successively added to the table plan — it might be better to proceed by *subtraction*; in other words, we start with all possible arrangements around the table, and subtract those where one or more couple is together. We will analyse the case of 6 couples, but we will be careful, in constructing our argument, that it works equally well for any value of n.

There are clearly $6! \times 5!$ ways to place the women and men alternately around a round table. Some of these arrangements will result in exactly no couple being together; that is what we aim to count. The others will involve between one and six couples being together. If we can count these, and then subtract the total from $6! \times 5!$, we will have achieved our objective. The focus has now shifted from individuals to couples, so let us denote these by $C_1, C_2, \cdots, C_6$.

Is it easy to count the number of ways in which C_1 is together, but no other couple? This would mean that the other five couples would fill the remaining seats with men and women alternating and no couple together. What about the number of ways in which C_1 and C_2 are together, but no others? That is not so easy, as it seems to split into two cases. One is when C_1 and C_2 are next to one another, and the other is when they are separated. We appear to be slipping into another unpleasant case analysis. How can this be avoided? What is easy to count?

After a little thought, we have an answer to this question. Surely it is possible to count the number of ways in which a particular couple is together, so long as we *do not care* what is happening to the others (so long as the men and women are still alternating around the table). The trouble with this, of course, is that there is a danger of counting things more than once.

Let A_i stand for the set of arrangements in which the couple C_i is together, irrespective of what the other couples are doing. To deal with the problem of overcounting, we can use the principle of inclusion and exclusion, abbreviated to PIE, which is described in [4, p60]. This approach *allows* over-counting to take place but *compensates* for it by making a sequence of adjustments.

We must introduce more notation. Let A_{ij} be the set of arrangements in which C_i and C_j are together (irrespective of what other couples are doing), A_{ijk} the set in which C_i, C_j and C_k are together, and so on. It is obvious that the sizes of these sets do not depend on which particular couples have been chosen; for example $|A_{135}| = |A_{234}| = |A_{256}|$.

Now PIE tells us that

$$\begin{aligned} m(6) = 6! \times 5! - \binom{6}{1}|A_1| + \binom{6}{2}|A_{12}| - \binom{6}{3}|A_{123}| \\ + \binom{6}{4}|A_{1234}| - \binom{6}{5}|A_{12345}| + |A_{123456}|. \end{aligned} \tag{1.1}$$

We need a consistent method for evaluating the sizes of these sets. We illustrate this by looking at A_{123}, the set of arrangements where couples C_1, C_2 and C_3 are sitting together. Remember that this method should be applicable whatever the total number of couples and whichever subset is chosen.

Begin by seating W_1. Now we have two choices for M_1 since he has to sit next to her. We then count round the table beginning with W_1 and then M_1, moving clockwise or anticlockwise as appropriate. The remaining ten seats are filled by the individuals W_4, W_5, W_6, M_4, M_5, M_6 and the two remaining couples C_2 and C_3. Since our list begins with W_1, M_1, the remaining individuals are in the order $W, M, W, M, \ldots$ regardless of the couples. The women can be chosen in order in 3! ways, as can the men, so there are $(3!)^2$ ways of placing the individuals.

Now we seat C_2 and C_3. They should go into gaps between the other diners (but not splitting C_1), and there is no reason why they should not both be in the same gap. The man and woman in each couple can be arranged so as to maintain the alternating pattern around the table. Since there are seven 'objects' and the table is round, there are seven gaps. So we begin by inserting two *markers* into the seven gaps, and we shall allow both markers to occupy the same gap.

If we were just choosing two different gaps from seven, there would be $\binom{7}{2}$ ways of doing so. This is a *combination without replacement*. In this case, however, we allow the same gap to be chosen twice. That makes it into a *combination with replacement*. This is a choice of r objects from n distinguishable objects, in which the order of selection does not matter and in which we allow objects to be chosen more than once. We will use $\left(\!\binom{n}{r}\!\right)$ to denote the number of ways of choosing r from n objects where replacement is permitted. In [4, p43], we showed that $\left(\!\binom{n}{r}\!\right) = \binom{n+r-1}{r}$.

The number of ways of choosing the gaps is $\left(\!\binom{7}{2}\!\right)$. Finally we *allocate* C_2 and C_3 to the markers, which can be done in 2! ways. It makes no difference whether or not the markers are in different gaps.

As a result of all this, we have $|A_{123}| = 2 \times 3! \times \left(\!\binom{7}{2}\!\right) \times 2! = 4032$.

Exactly the same argument applies to the following calculations.

$$
\begin{aligned}
|A_1| &= 2 \times (5!)^2 \times \left(\!\binom{11}{0}\!\right) \times 0! = 28\,800;\\
|A_{12}| &= 2 \times (4!)^2 \times \left(\!\binom{9}{1}\!\right) \times 1! = 10\,368;\\
|A_{1234}| &= 2 \times (2!)^2 \times \left(\!\binom{5}{3}\!\right) \times 3! = 1680;\\
|A_{12345}| &= 2 \times (1!)^2 \times \left(\!\binom{3}{4}\!\right) \times 4! = 720;\\
|A_{123456}| &= 2 \times (0!)^2 \times \left(\!\binom{1}{5}\!\right) \times 5! = 240.
\end{aligned}
$$

In the preceding list, there are some 'redundant' expressions such as 0!; these were included to emphasize the shape of the formulae.

When we feed all of this into (1.1), we obtain the expression

$$m(6) = 6! \times 5! - \binom{6}{1} \times 28\,800 + \binom{6}{2} \times 10\,368 - \binom{6}{3} \times 4032$$
$$+ \binom{6}{4} \times 1680 - \binom{6}{5} \times 720 + \binom{6}{6} \times 240$$

which is 9600, and we have the solution to problem 1.1.

That was hard work, and you might be disappointed at the form of the solution. The outcome of the analysis is a summation, and it turns out to be impossible to reduce it to a simpler expression. If you are familiar with derangements, you will already have met an instance where this happens. It is still a solution to the problem posed.

We now need to work on producing a general formula, but let us look at the case $n = 6$ and try to produce a formula for the number of arrangements with r particular couples together, where $1 \leq r \leq 6$. Take, as an example, $A_{1234} = 2 \times (2!)^2 \times \left(\!\binom{5}{3}\!\right) \times 3!$ and so $r = 4$. The 2! refers to the individuals, which in this case was $(6-4)!$ and so would be $(6-r)!$ in general. The combination with replacement refers to the number of gaps when the individuals have been placed and to the number of markers to enter. Now there are $2(6-r)$ individuals, together with couple C_1, so there are $13 - 2r$ gaps. We must place $r-1$ markers, so the combination in question is $\left(\!\binom{13-2r}{r-1}\!\right)$ which can be simplified to $\binom{11-r}{r-1}$.

Finally, the 3! refers to the allocation of the remaining couples, and this becomes $(r-1)!$. Hence the formula we need is

$$|A_{(r)}| = 2 \times (6-r)!^2 \binom{11-r}{r-1} \times (r-1)!$$

where the subscript (r) indicates that this is for a set of r particular couples sat together. A further simplification is possible, using the identity $\binom{11-r}{r-1} \times (r-1)! = \frac{(11-r)!}{(12-2r)!}$, which you can check algebraically.

Incorporating this into (1.1), we obtain

$$m(6) = 6! \times 5! + 2\sum_{r=1}^{6}(-1)^r \binom{6}{r} (6-r)!^2 \frac{(11-r)!}{(12-2r)!}.$$

In fact, if we substitute $r = 0$ into the expression inside the sum, it turns out, rather conveniently, to be $6! \times 5!$ (which is not really an accident, since we are counting the arrangements with at least no couples together). Hence we can simplify the formula further, and then generalise it to any value of n to obtain a final formula

$$m(n) = 2\sum_{r=0}^{n}(-1)^r \binom{n}{r}(n-r)!^2 \frac{(2n-r-1)!}{(2n-2r)!}.$$

The values of $m(n)$ for the first few values of n are given in the following table. It is reassuring to see that our formula gives $m(5) = 312$, the answer we found at the start of the section.

n	3	4	5	6	7	8	9
$m(n)$	2	12	312	9600	416 880	23 879 520	1 749 363 840

Table 1.1

In 1891 Édouard Lucas named this as the *problème des ménages* and this is how it is known today. This, incidentally, accounts for the choice of the notation $m(n)$. There are various versions of the problem, one of which does not employ the round table convention but makes the seats distinguishable. This interpretation produces the sequence $2n \times m(n)$ and the value 115 200 when $n = 6$.

An alternative solution to this problem, using transparent dominos, can be found in [3].

Exercise 1a

In this exercise, you are asked to carry out similar calculations in situations where the conditions of the ménage problem are changed slightly. In each case you should derive a general expression for n couples as well as the solution for six couples.

1. Find the number of ways in which six married couples can be seated in a row on a bench (without the requirement that men and women alternate) so that

   (a) no wife is to the immediate right of her spouse;

(b) no husband and wife are sitting next to one another;

(c) no wife is sitting anywhere to the right of her spouse.

2. Find the number of ways in which six married couples can be seated in a row on a bench with men and women alternating so that
   (a) no wife is to the immediate right of her spouse;

   (b) no husband and wife are next to one another;

   (c) no wife is sitting anywhere to the right of her spouse.

3. Find the number of ways in which six married couples can be arranged around a circular table (without the requirement that men and women alternate) so that no husband is diametrically opposite his wife.

1.2 Matrimonial affairs

It is a truth universally acknowledged, that a single man in possession of a good fortune, must be in want of a wife.

Much of the social life of Regency England revolved around recognition of this fact. Young gentlemen hoping to make their way in society would do their best to meet eligible young ladies.

Conversely, the major task of any self-respecting matriarch was to see that her daughter made an advantageous marriage before reaching the age of twenty-five.

Imagine, then, the situation of six local beaux, whose names are Brandon, Churchill, Darcy, Ferrars, Knightley and Wentworth, attending a county ball at which the debutantes Anne, Emma, Harriet, Isabella, Jane, Lydia, Mary and Sophia are being presented. Imagine that an exit poll, asking each of the young gentlemen which of the young ladies he would countenance as a wife, yields the following results. In table 1.2, the gentlemen are indicated by the capital initials of their surnames and the ladies by the lower-case initials of their first names.

Gentlemen	Ladies
B	$\{a, e, h, i\}$
C	$\{i, j, l\}$
D	$\{j, m\}$
F	$\{m, s\}$
K	$\{e, m, s\}$
W	$\{h, l\}$

Table 1.2

Problem 1.2

Is it possible to pair up the six eligible bachelors with six young ladies in a way which takes male partialities into account?

Note that the girls are not consulted in this procedure, and that polygamy is not *comme il faut* in Nineteenth Century England. A little experiment shows that this is quite easy to achieve. One solution is to pair the partners as follows

$$(B, a)(C, i)(D, j)(F, m)(K, e)(W, h)$$

where each man has been assigned the first name alphabetically in his list. It is clear by experimentation that there are many other ways of doing this. It is worth noting, however, that if you began with the pairs (B, h) and (C, l), you would be in trouble, since poor Wentworth would be deprived of an acceptable spouse. It would seem, all the same, that it is not too hard to avoid such a catastrophe; an informal strategy might be to begin with the most fastidious of the beaux, Darcy, Ferrar and Wentworth, and leave the easy-going Brandon until last.

Now suppose that the poll had resulted in a quite different set of choices, such as the one shown in table 1.3.

Gentlemen	Ladies
B	$\{e, f, j, l\}$
C	$\{a, h, i, m, s\}$
D	$\{e, j\}$
F	$\{e, f, l\}$
K	$\{f, j\}$
W	$\{j\}$

Table 1.3

No young man is ungallant enough to list no preferences, and no deb is an obvious wallflower, but it still turns out to be impossible to satisfy everybody, even if we adopt the strategy of 'fussiest first'. It is clear that (W, j) must be one of the pairs, and then D and K must settle for e and f respectively. This leaves F with l and now B, despite spreading his affections widely, has no available partner.

We discovered this impossibility by focusing on a subset $\{B, D, F, K, W\}$ of the gentlemen, and noticing that the set of ladies they favour is $\{e, f, j, l\}$. It is clear by common sense, or the pigeonhole principle, that these two sets cannot be matched.

Hence a matching is impossible if there is any set of m gentlemen who have collectively nominated fewer than m ladies. It follows that a necessary condition for a matching to exist is that every set of m gentlemen together has nominated at least m ladies. This must be true for every m in the range $1 \leq m \leq N$ where N is the number of beaux at the ball.

We will call the set of ladies nominated collectively by a given set of gentlemen their *pool*, and we shall introduce some formal notation so that we can talk about what is going on. Denote the set of gentlemen by G and the set of ladies by L. For $g \in G$, let L_g be the set of ladies on his list. Now, for any $S \subseteq G$, the pool associated with S is $L_S = \bigcup_{g \in S} L_g$. Our necessary condition for a matching now becomes

$$|S| \leq |L_S| \text{ for all } S \subseteq G. \tag{1.2}$$

We shall call this *Hall's condition*; you will see the reason for this later. The obvious question to ask is whether Hall's condition is *sufficient*. In other words, if we know that it is true, does it follow that a matching must exist?

Let us examine this condition for some small values of $|S|$.

Suppose first that Hall's condition is true but only for subsets with one element — that is, whenever $|S| = 1$. This means that there is no curmudgeon among the young men who is not willing to countenance any of the ladies as his wife. This is a good start, and if there is only one gentleman in G, it certainly means that he will be able to claim his partner. However, there is no guarantee that two gentlemen will both be happy, since they might both have named the same debutante and no other. It is clear that this condition for single element subsets will not guarantee a match when $N > 1$.

If, on the other hand, Hall's condition is true only for subsets with two elements — when $S = 2$ — then every pair of beaux will nominate at least two ladies between them. Does this mean that both of them will be lucky in love? But now it is possible that the first of these gentlemen has nominated two ladies, whereas the second has listed none. It follows we are not going to get anywhere unless the condition is true when $|S| \leq 2$. Suppose now that there are exactly two gentlemen whose pool consists of two ladies, and that they have each nominated at least one. Then either they are both indifferent to which lady they marry (in which case there are four ways of matching), or one of them is fussy and only desires a particular lady, in which case the other has to settle for the second, or they are both fussy but desire different ladies, and that is quite convenient.

So it appears that (1.2) holds for all sets of size 1 and 2 is sufficient so long as there are exactly two gentlemen involved. Add a third gentleman, however, and this is not enough, since all three gentlemen might make the same choice of two ladies. One way of preventing this happening is to require that Hall's condition is true when $|S| = 3$.

We can carry on doing this for a while and it will seem more and more likely that Hall's condition, applied to all possible subsets, is sufficient to guarantee a matching. However, the argument is going to get more involved as the number of gentlemen increases. All the same, we have at our disposal a proof technique which is tailor-made for dealing with statements about all positive integers — namely *mathematical induction*. The induction hypothesis is the obvious one.

If $|G| = n$ and $|S| \leq |L_S|$ for all $S \subseteq G$, then there is a matching.

We are going to use induction on n. In fact, the arguments we have made above concerning small sets suggest that we should use complete induction. We assume that the statement is true for all sets of size less than k and show that it is also true for sets of size k. This, together with the usual base case, shows that it is true for all sets.

Base

The statement is obviously true for $|G| = 1$.

Step

Assume that the statement is true for all sets G with $|G| < k$.

In order to clarify the exposition, we will focus on a particular value of k, which might as well be 50. Clearly the argument we produce should not be dependent on this number, and we should be able to formulate it in the general case.

Let us take a particular set G of fifty gentlemen and assume that $|S| \leq |L_S|$ for all $S \subseteq G$. The induction hypothesis tells us that we have a matching for any proper subset of G, and our aim is to produce a matching for G itself.

Suppose we know that, for any proper subset S, $|L_S| \geq |S| + 1$. This is a generous situation since it says that each individual man has listed at least two ladies, every pair have chosen at least three, and so on up to any set of 49 gentlemen whose pool is at least 50. We now select an arbitrary gentlemen, X, with an arbitrary lady on his list, x, and ask them to sit down together at the side of the ballroom. We then remove x from all the other lists. We now have 49 gentlemen remaining, and the pool for any subset of size k contains at least k girls. (The 'worst-case scenario' would be that everyone had chosen lady x as part of their list). But we know that a matching is now possible for these 49 gentlemen, and so, asking X and x to stand up again, we have managed to match all 50 of them.

We have now dealt with the case where the induction step is easy. What about the awkward case? What does this mean? It means that there is some subset of k gentlemen (for some $1 \leq k \leq 49$) whose pool contains *exactly* k ladies. Let us suppose, for the sake of argument, that the k in question is 23. Since this is less than 50, we know that these 23 men can be matched with 23 ladies of their choice. Again, we ask these couples to move to the side of the dance floor so that we can concentrate on the remaining 27 gentlemen, and we delete the names of the chosen ladies from their lists.

Now think about a subset S of these 27 gentlemen, with, say, $|S| = 11$. What can we say about the number of ladies in the *modified* L_S? Could this be fewer than 11? Think about these 11 gentlemen and the 23 others who are watching from the sidelines. We know that the latter had exactly 23 ladies in their pool (who are not in the modified L_S), so if it were the case that $L_S < 11$, we would now have a group of 34 gentlemen whose pool contains fewer than 34 ladies. However, we know that is not the case since this would violate Hall's condition. Therefore $|L_S| \geq 11$, and so for this set $|S| \leq |L_S|$. But there was nothing special about this subset of the 27 gentlemen; the same is true for any subset. Hence, by induction, we have a matching for these gentlemen. Now we invite the 23 couples back onto the dance floor, and we have a matching for all 50 gentlemen as required.

This proves the inductive step. Hence, by induction, the statement is true for sets of any size.

This is a subtle argument (which is attributed to Halmos and Vaughan in [2]) and it may be worth rereading it carefully. Of course, there is nothing special about the numbers 50, 27, 23 and 11, and these could be replaced by any numbers. You should try to construct this induction argument in general terms.

Hence the question asked can be answered in the affirmative, as follows:

Hall's marriage theorem

Each of N young men can be matched with a young lady from his list of preferences if, and only if, for each m such that $1 \leq m \leq N$, the pool for every subset of m men contains at least m ladies.

This result is due to Philip Hall, who proved it in 1935. Its import goes beyond the politics of marriage-brokering. Obviously the actors could be changed from Regency beaux to friends sharing a box of chocolates with everyone writing down which ones are acceptable. At a more prosaic level, we could have a group of employees in a firm who have different skills: is it possible for jobs to be allocated in such a way that everyone does something in which they are proficient? In the US, there is a service called the National Resident Matching Program, which attempts to ensure that every medical student applying for training is able to attend their college of choice. This (under what are probably different conditions) makes essential use of algorithms related to the marriage theorem.

The marriage theorem can be stated in a way which does not refer to beaux or debutantes in the language of set theory:

HALL'S MARRIAGE THEOREM: TRANSVERSAL FORM

Suppose we have n subsets $S_1, S_2, \ldots, S_n$ of a finite set X. We call a subset T of X a *transversal* if it consists of n distinct elements $\{s_1, s_2, \ldots, s_n\}$ such that $s_i \in S_i$ for $1 \le i \le n$. A transversal exists if, and only if, for every $I \subseteq \{1, 2, \ldots, n\}$, $| \bigcup_{i \in I} S_i | \geq | I |$.

In the context of the county ball, X is the set of debutantes, S_i is the list favoured by gentleman i and the transversal is the successful allocation of ladies to gentlemen. The role of I in the statement of the theorem is to ensure that every possible subset of gentlemen is tested. If there are m numbers in the set I, then $\bigcup_{i \in I} S_i$ represents their pool, and the condition states that it is of size m or more.

One could also show the process in a table with rows and columns. The rows would be the elements of the set X (the ladies) and each row would correspond to a different i (or gentleman) and would indicate his choice of ladies (the set S_i). A transversal would consist of an element chosen from each row of the table so that no two elements were in the same column.

1.3 Latin squares

In the dining hall of Gonville and Caius College in Cambridge, several of the stained glass windows commemorate former fellows. Some of these are mathematicians; they include John Venn, who popularised the set-theoretical diagrams which appear later in this chapter, George Green, who worked in electromagnetism, and the statistician R. A. Fisher. They were designed by a current fellow, A. W. F. Edwards, yet another mathematician.

The window commemorating Fisher is a square array consisting of 49 panes of coloured glass. Each colour appears once in each row and once in each column of the array. Representing the different colours by the numbers 1 to 7, Fisher's window is shown in figure 1.4.

1	2	3	4	5	6	7
2	6	1	7	3	5	4
6	5	7	2	4	1	3
5	4	6	3	2	7	1
4	7	2	6	1	3	5
3	1	4	5	7	2	6
7	3	5	1	6	4	2

Figure 1.4

Such an array is known as a *Latin square* of order 7. In general, a Latin square of order n is an $n \times n$ array, with entries in the set $\{1, 2, \ldots, n\}$, in which each entry occurs exactly once in each row and exactly once in each column. There is a vast literature on how to construct Latin squares and their properties, but we will restrict ourselves to the following problem.

Problem 1.3

How do we build a Latin square row by row?

Let us be clear what this means in a general context. Suppose we are trying to create a Latin square of order n. It is easy to write down a first row; it is simply the numbers 1 to n in some order. It is also pretty obvious that, whichever first row is chosen, it is possible to create a second row, again with the numbers 1 to n, in which none of the n columns contains a repeated number. Again, suppose that this is selected at random. Is it possible to keep on doing this, always ensuring that no column contains a repeat, and finish up with a Latin square?

In other words, we are not concerned about using an algorithm for creating Latin squares; there is an obvious one which involves just beginning with the numbers in order and then cycling them one by one each row. We are asking if *any* process, in which the resulting rectangular arrays do not obviously contravene the 'Latin property', is going to succeed.

The first row is arbitrary, but without loss of generality, we will start with the numbers in the usual order.

1	2	3	4	5	6	7

Figure 1.5

Next we add a second row as shown in figure 1.6. The only criterion is that no number replicates the one in the first row. Following Edwards, we make the following choices.

1	2	3	4	5	6	7
2	6	1	7	3	5	4

Figure 1.6

For the third row, we again wish to conserve the Latin square property, so we have to assign the numbers 1 to 7 so that none of them appears above it in the same column. Again we adopt the choices made in the window.

1	2	3	4	5	6	7
2	6	1	7	3	5	4
6	5	7	2	4	1	3

Figure 1.7

At this point, we pause for thought. Things are going like clockwork, but is this approach always going to work? Surely it is possible — indeed, perhaps probable — that we will run out of options?

Let us introduce a little bit more terminology. We call the array we have reached a *Latin rectangle*. Specifically, it is a 3×7 Latin rectangle; we have three rows, seven columns and seven numbers to choose from. So can this be developed into a 4×7 Latin rectangle? And can the argument we use for this generalise so that we will eventually produce a 7×7 Latin square?

Let us analyse exactly what we are required to do. We have the following Latin rectangle.

1	2	3	4	5	6	7
2	6	1	7	3	5	4
6	5	7	2	4	1	3

Figure 1.8

Now we have restricted choices as to what we can do next. Column by column, we are restricted to choosing a number from each of the following sets.

$\{3,4,5,7\}$ $\{1,3,4,7\}$ $\{2,4,5,6\}$ $\{1,3,5,6\}$ $\{1,2,6,7\}$ $\{2,3,5,7\}$ $\{1,2,5,6\}$

It now becomes clear that we are back in the land of Hall's theorem. We are looking for a transversal from these seven sets of four elements. Hence all we have to do is to decide if any union of k of these sets contains at least k elements. This, however, is not as easy as it looks; there is a great deal of checking to do. However, let us examine this decision calmly. We notice that

- each set contains exactly 4 elements, since the entries are different in each of the existing three rows;
- each element, from 1 to 7, occurs in exactly 4 of the sets; this is because each is entered in a different column in each of four rows.

Moreover, it is worth observing that these facts about the sets will be true at every stage in the process. After just one row, the sets each contained 6 elements and each element appeared in exactly 6 of the sets, since it was only excluded in one of the columns. In fact, the observations are obvious since the only exclusions are the numbers appearing in each column of the Latin rectangle above, and each element is being treated symmetrically.

So let us now consider what happens if we form the union of four of the sets. This will produce 16 elements, but this will include repetitions. To take an example, suppose we selected the sets in columns 1, 3, 4 and 7; we obtain the list $\{1,1,2,2,3,3,4,4,5,5,5,5,6,6,6,7\}$. No element can be in this 'set' more than 4 times, so (by the pigeonhole principle) the list must contain at least four different elements. (In our example, there are actually seven different elements, so this is a very rough lower bound.) This is exactly the condition for a transversal to exist.

Hall's theorem therefore shows that this intuitive approach to building a Latin square works. In exercise 1b, you are asked to show that this argument works in a completely general context.

Latin squares are used in the design of statistical experiments, and Fisher's book written in 1935 is a seminal work on this topic. Let us give an example of this.

Problem 1.4

An agriculturalist wishes to measure the effect of seven different fertilisers on the yields of a particular crop. Tests are to be carried out by planting the crop in a square field with a road along its Southern boundary and a river along its Western boundary. It is possible that distances from the road and the river, as well as the type of fertiliser used, may affect crop yields. Is it possible to design an experiment which contrasts the effects of the different fertilisers on crop yield while eliminating the influence of the road and river?

There are many factors which affect crop yield, such as soil acidity, humidity and amount of sunlight, but in this simplified example we are assuming that there are only three which matter — the type of fertiliser used, the distance from the road and the distance from the river. Clearly we are going to divide the field up into regions, plant them with seed in exactly the same way, and then allocate the fertilisers in a systematic way. We want the regions which have been treated with any pair of fertilisers to be affected in the same way by road and river distances. To be precise, we would like the regions to have the same average distances to the road and to the river.

An obvious way of doing this is to divide the field into a square grid of 49 regions. We then allocate the numbers 1 to 7 to the regions, representing the fertilisers used, in such a way that

- every number appears exactly once in each row (to mitigate the road influence);
- every number appears exactly once in each column (to mitigate the river influence).

So what is needed is a Latin square of order 7. It is this which is being celebrated in the Fisher window.

1.4 Three sisters

Those of you who are familiar with polite society will be aware that Jane, Lydia and Mary are sisters, and, moreover, that their mother is a lady who does not take no for an answer. This is particularly so when it involves her daughters, so it turns out that we must refine the marriage problem.

Problem 1.5

With the list of preferences from the local beaux below, is it possible to find a matching which includes Jane, Lydia and Mary?

Gentlemen	Ladies
B	$\{a, e, h, i\}$
C	$\{i, j, l\}$
D	$\{j, m\}$
F	$\{m, s\}$
K	$\{e, m, s\}$
W	$\{j\}$

Table 1.4

First it is clear that W, being rather pernickety, has to pair with j. Next it follows that we must have the pairs $(D, m), (F, s)$ and (K, e). Finally B and C can choose freely from the other debs; we have a matching and a happy mother.

On the other hand, there are clearly lists which permit a matching but do not accommodate all the sisters. A trivial one is where each gentleman is only prepared to countenance one lady. If these ladies are all different, there is clearly a matching, but since there are more ladies than gentlemen it is possible that Mary is not paired with any gentleman.

Table 1.5 shows a less obvious example

Gentlemen	Ladies
B	$\{a, e, h, i\}$
C	$\{e, i, s\}$
D	$\{a, e, s\}$
F	$\{a, j, l, m, s\}$
K	$\{i, e, s\}$
W	$\{e, i, j, l, m\}$

Table 1.5

The gentlemen here are not being at all fussy and there many possible matchings. Unfortunately, the pool for B, C, D and K is $\{a, e, h, i, s\}$, which does not contain any of the three sisters. Hence they must be allocated to F and W, and, despite the fact that both these gentlemen would welcome any of the sisters, this is impossible without bigamy.

So we have a very clear criterion here for preventing a matching which includes the three sisters. If we can find a pool of four gentlemen which omits three of the sisters, or a pool of five gentlemen which omits two of them, we will not be able to match all of them. Turning this idea round (and generalising it) we clearly have the following *necessary* condition for a matching which accommodates a group of sisters:

Hall's condition with sisters

Suppose there are N gentlemen and a group of sisters amongst the ladies. For any k gentlemen whose pool omits r of the sisters, we need $r \leq N - k$.

Now we ask the same question as we did in the discussion of the marriage theorem. Is this condition (together with Hall's condition which guarantees a matching) also *sufficient*? It might seem surprising that such a weak condition is adequate, but it turns out to be the case that it is.

In the general case we have a set $G = \{g_1, g_2, \ldots, g_N\}$ of gentlemen and a set $L = \{l_1, l_2, \ldots, l_M\}$ of ladies, with $N \leq M$. We also have a set $S = \{l_1, l_2, \ldots, l_m\}$ of sisters for some $m \leq M$. We have placed the sisters at the head of the set listing for L, but there is no loss of generality in doing so.

We have two conditions for the existence of a matching which includes the set S of sisters:

(i) the pool for each set of k gentlemen contains at least k ladies;

(ii) for any set of k gentlemen, there are at most $N-k$ sisters not in their pool.

The proof of sufficiency will be left as an exercise for the reader. The key idea in the argument combines the virtues of both democracy and élitism. We invent $M-N$ extra gentlemen, thereby achieving equal numbers of either sex. However, these *bogus* gentlemen are definitely not upper crust, and they are each (without consultation) allocated the list of all the ladies apart from the sisters . What is so ingenious about this is that if we are able to find a matching of all the (real and bogus) gentlemen, we can be sure that sisters all have acceptable partners.

This result can also be expressed in set-theoretical language as follows:

Hall's theorem with sisters: transversal form

Suppose we have $n+1$ subsets $S_1, S_2, \ldots, S_n$ and Y of a finite set X. We call a subset T of X a transversal if it consists of n distinct elements $\{s_1, s_2, \ldots, s_n\}$ such that $s_i \in S_i$ for $1 \leq i \leq n$. A transversal which includes all the elements of Y exists if, and only if, we have

(i) for every $I \subseteq \{1, 2, \ldots, n\}, \mid \bigcup_{i \in I} S_i \mid \geq \mid I \mid$;

(ii) for every $I \supseteq \{1, 2, \ldots, n\}, \mid Y \cap \bigcap_{i \in I} S_i^C \mid \leq n- \mid I \mid$.

Exercise 1b

1. Consider the following preference lists. Is a matching possible? If not, explain why it is impossible. (As before, the capital letters represent the gentlemen, and the lower case letters represent the ladies.)

(a)	B	C	D	F	G	W
	a,e	e,h	h,s	i,s	i,j	a,j
(b)	B	C	D	F	G	W
	a,e,h,i	e,h,i,j	a,h,i,j	a,e,i	a,e,h	a,e,i,j
(c)	B	C	D	F	G	W
	a,h	a,e	e,h	l,m	l,s	m,s
(d)	B	C	D	F	G	W
	a,h,m,s	a,h,m,s	a,h,m,s	a,h,m,s	e,j,l,m	e,j,l,m
(e)	B	C	D	F	G	W
	a,e	a,i,s	a,e,s	a,s	a,h,s	a,e,h

2. Consider the following preference lists. In each case, is it possible to find a matching which includes the three sisters, j, l and m? If not, explain why.

(a)	B	C	D	F	G	W
	a,e,j	h,j,m	a,j	e,l	a,e,m	e,m
(b)	B	C	D	F	G	W
	a,e,j	j,l,m	a,j	l,m	a,e,m	j,m
(c)	B	C	D	F	G	W
	a,e,m	j,l,s	i,m	e,h	a,e,h,i	e,i,m

3. Prove the sufficiency of the condition for the existence of a matching which accommodates a privileged group of ladies.

4. A pack of 52 playing cards is shuffled and dealt into 13 piles of four cards. Show that it is possible to select one card from each pile to get one card of each denomination.

5. There are N boys and N girls at a party. If every subset of k boys collectively knows at least k girls, prove that every subset of k girls knows at least k boys. (You may assume that 'knowing' is a symmetric relationship: if a knows b then b knows a.)

6. A tennis tournament involving $2n$ players is held over $2n - 1$ days. Every day there are n matches each of which produces a winner and a loser. Over the course of the tournament, every player meets every other player exactly once. Is it possible to pick a winning player each day without picking the same player more than once?
[Putnam 2012, adapted]

7. A $2n$ by $2n$ chessboard has n rooks in every row and n rooks in every column. Show that there exist $2n$ rooks none of which share a row or a column.

8. Prove that an $r \times n$ Latin rectangle (with $r < n$) can be extended to an $n \times n$ Latin square.

9. Is it possible to extend the Latin 3×5 rectangle below to a 7×7 Latin square?

1	2	4	6	7
2	4	1	7	5
5	7	6	2	4

1.5 Walking down the street

Problem 1.6

Fifteen schoolgirls walk to church every day for a week in five rows of three. Is it possible to organise this so that no two girls ever walk abreast on more than one day?

This problem was posed in 1850 by the Reverend Thomas Kirkman, an amateur mathematician of some standing, in *The Lady's and Gentleman's Diary*, a recreational mathematics magazine for the gentry. It ceased publication in 1871, and unfortunately neither *Country Life* nor *OK!* has decided to restart the tradition. It was common in Victorian times to call such a procession a *crocodile*, despite the anatomical incorrectness of this image. Perhaps a centipede would be more accurate, but we shall retain the terminology, since it is a convenient way of referring to a 'long line of people, especially school children, moving in file' (as the dictionary definition has it).

There can be little doubt that the magazine did not require solvers to provide a mathematical analysis of the problem, expecting instead a 'trial-and-error' approach with something of the flavour of a Sudoku puzzle. Actually none of the subscribers succeeded in doing this, but Kirkman himself had several solutions. Let us, therefore, enter into the spirit of the enterprise and have a go.

Allow the girls to form the first day's crocodile for themselves, and then label them using the numbers 1 to 15. Represent the first day's crocodile as shown in figure 1.9.

$$\begin{array}{c} \{1,2,3\} \\ \{4,5,6\} \\ \{7,8,9\} \\ \{10,11,12\} \\ \{13,14,15\} \end{array}$$

Figure 1.9

Now we try to build up the second crocodile. We might as well begin with girl number 1. Neither 2 nor 3 will feature in the first triple, so the second schoolgirl might as well be 4. Similarly, the third will be 7, and we have the first triple on the second day. Using the same logic, it might seem sensible to choose $\{2,5,8\}$ and $\{3,6,9\}$ for the next two triples as shown in figure 1.10.

$$\begin{array}{cc} \{1,2,3\} & \{1,4,7\} \\ \{4,5,6\} & \{2,5,8\} \\ \{7,8,9\} & \{3,6,9\} \\ \{10,11,12\} & \\ \{13,14,15\} & \end{array}$$

Figure 1.10

This would leave us having to organise 10 to 15 into two triples, but a moment's thought should convince us that this is impossible without repeating two pairs. It is more sensible to anticipate this problem and choose $\{2,5,10\}$ and $\{3,6,13\}$ as the second and third triples as in figure 1.11.

$$\begin{array}{cc} \{1,2,3\} & \{1,4,7\} \\ \{4,5,6\} & \{2,5,10\} \\ \{7,8,9\} & \{3,6,13\} \\ \{10,11,12\} & \\ \{13,14,15\} & \end{array}$$

Figure 1.11

Now we must allocate the remaining schoolgirls — 8, 9, 11, 12, 14 and 15 — between the last two triples, and this can be done as shown in figure 1.12.

$$\begin{array}{cc} \{1,2,3\} & \{1,4,7\} \\ \{4,5,6\} & \{2,5,10\} \\ \{7,8,9\} & \{3,6,13\} \\ \{10,11,12\} & \{8,11,14\} \\ \{13,14,15\} & \{9,12,15\} \end{array}$$

Figure 1.12

You are invited to complete this activity in exercise 1c, but do not worry if you cannot manage it; the original problem was not solved by any of the readers of the *Diary*. It turns out that there are exactly seven different solutions.

It is worth asking why there are fifteen schoolgirls and seven days. Can this be extended from a week to a fortnight? As already stated, there are 105 different pairs. Each triple uses up three pairs, so in every daily crocodile there are fifteen. Hence the most we can ask for is a week of arrangements without repeating a pair. This means that, if you propose a solution, it is easy (if time-consuming) to check whether it is valid; you can simply tick off the different pairs of schoolgirls, and see if there are the right number of them.

We attempt a more systematic analysis. It is usually a good idea to begin by considering smaller numbers. The simplest case is when there are four girls, when a crocodile would consist of two pairs — walking four in a row or in single file would be antisocial or eccentric. As there are six pairs altogether, we ask whether the girls, whom we number 1 to 4, can walk to church for three days without repeating a pair. This turns out to be possible, and a solution is shown in figure 1.13.

$$\begin{array}{ccc} \{1,2\} & \{1,3\} & \{1,4\} \\ \{3,4\} & \{2,4\} & \{2,3\} \end{array}$$

Figure 1.13

Moreover, all solutions to this problem are effectively the same. Each crocodile is determined by 1's partner, and there are only three choices.

Now let us consider eight girls, who can form 28 pairs. (You will be asked to consider six girls in exercise 1c.) The girls are numbered from 1 to 8. The first crocodile can be formed as shown in figure 1.14.

$$\begin{array}{c} \{1,8\} \\ \{2,7\} \\ \{3,6\} \\ \{4,5\} \end{array}$$

Figure 1.14

Now it is tempting to cycle the girls around, producing the sequence of crocodiles shown in figure 1.15.

$$\begin{array}{cccc} \{8,7\} & \{7,6\} & \{6,5\} & \{5,4\} \\ \{1,6\} & \{8,5\} & \{7,4\} & \{6,3\} \\ \{2,5\} & \{1,4\} & \{8,3\} & \{7,2\} \\ \{3,4\} & \{2,3\} & \{1,2\} & \{8,1\} \end{array}$$

Figure 1.15

The first three crocodiles each consist of four new pairs, but the final one involves the same four pairs as we started with. A way of avoiding this unfortunate duplication is to fix one of the girls, who then functions as a sort of *pivot* around which the others revolve. If 8 plays this role, we obtain the sequence of crocodiles shown in figure 1.16.

$\{1,8\}$	$\{7,8\}$	$\{6,8\}$	$\{5,8\}$	$\{4,8\}$	$\{3,8\}$	$\{2,8\}$
$\{2,7\}$	$\{1,6\}$	$\{7,5\}$	$\{6,4\}$	$\{5,3\}$	$\{4,2\}$	$\{3,1\}$
$\{3,6\}$	$\{2,5\}$	$\{1,4\}$	$\{7,3\}$	$\{6,2\}$	$\{5,1\}$	$\{4,7\}$
$\{4,5\}$	$\{3,4\}$	$\{2,3\}$	$\{1,2\}$	$\{7,1\}$	$\{6,7\}$	$\{5,6\}$

Figure 1.16

We can check that all 28 pairs have been included as follows. In each row, the sums for each pair are the same (working modulo 7) for each crocodile, and different for different crocodiles. So no pair appears in two crocodiles.

In fact, this method will work for any crocodile involving $2n$ girls walking in pairs. Of course, this is not just about schoolgirls. It concerns the very practical problem of scheduling a round robin tournament. The $2n$ objects are teams in a particular sport, such as football, where each team has to play every other exactly once in the tournament. We have shown that this can be done efficiently in n weeks. If, as in the Football League, teams are required to meet each other twice, once at home and once away, that can be accommodated by repeating the solution but switching the order in each pair. If there were an odd number of teams, then a schedule can be found by adding a dummy team, solving for the even case and then regarding a match against the dummy team as a rest day. There are many variations and it is possible to incorporate other constraints, but the problem is essentially solved.

Exercise 1c

1. Try to complete the process for finding a solution to Kirkman's schoolgirl problem started in section 1.5.

2. (a) Use the pivoting method to obtain a solution to the problem for six schoolgirls walking in pairs. Number the girls from 1 to 6 and use 6 as the pivot.

(b) The sequence of crocodiles shown in table 1.6 is a solution which does not appear to use pivoting. Can this solution be obtained from that in (a) by renumbering the schoolgirls?

{1,2}	{1,3}	{1,4}	{1,5}	{1,6}
{3,4}	{2,5}	{2,6}	{2,4}	{2,3}
{5,6}	{4,6}	{3,5}	{3,6}	{4,5}

Table 1.6

1.6 Finite geometries

Problems such as these can be represented geometrically. Take the simplest case of four schoolgirls. The pivoting method produces the result shown in figure 1.17.

{1,4}	{3,4}	{2,4}
{2,3}	{1,2}	{3,1}

Figure 1.17

We can consider the girls as vertices of a square. Pairs are then represented by lines joining two of the vertices. They are the four sides of the square and the two diagonals.

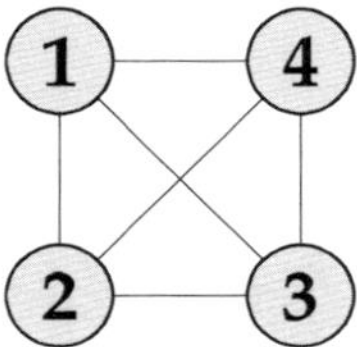

Figure 1.18

Now the three crocodiles needed are represented by choosing pairs of non-intersecting lines, as shown in figure 1.19.

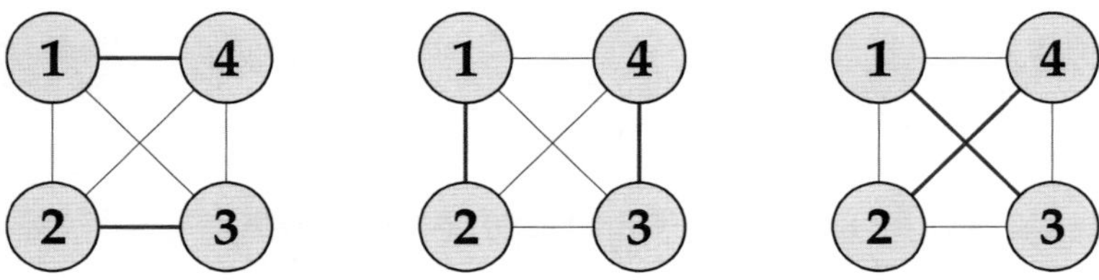

Figure 1.19

You will notice that the previous comment, despite its simplicity, appears to be incorrect on two counts. First, the 'lines' are actually segments, and, secondly, the diagonals clearly do intersect at the point in the centre. However, we have deliberately *redefined* our geometric terms. The two lines 13 and 24 do not meet since there is no point in the middle. This configuration is known as a *finite* geometry consisting of four points and four lines. Lines which do not meet are *parallel* so not only are 12 and 34 parallel, as are 14 and 23, but so are 13 and 24. Hence a crocodile is equivalent to two parallel lines.

Note also that each point in this finite geometry lies on exactly three lines, and each pair of lines which are not parallel meet at exactly one point.

When $n = 6$ the points form a regular hexagon, as in the diagram below. There are six points and fifteen lines. Every point lies on five lines and every line has two points. Two non-parallel lines intersect in one point and two distinct points lie on one line.

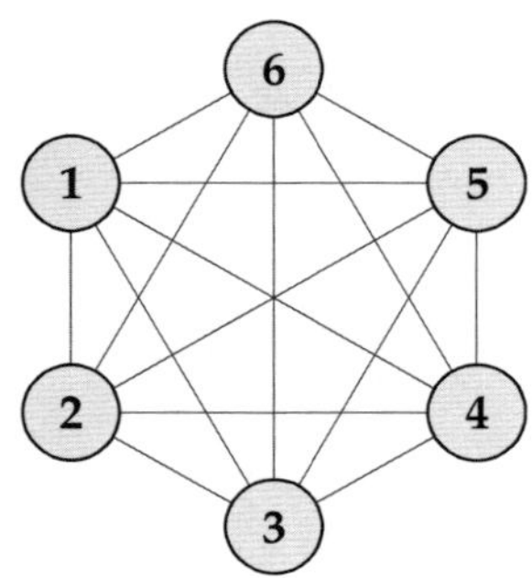

Figure 1.20

We can obtain a crocodile by using the three parallel (that is, non-intersecting) lines.

For example, we may go round the perimeter of the hexagon taking every other edge to obtain a crocodile in two different ways. This leaves us with the three diameters of the hexagon and six other diagonals to divide among three crocodiles. To do this we can choose each diameter together with the two diagonals which appear to be perpendicular to it in figure 1.20.

The result is shown in figure 1.21.

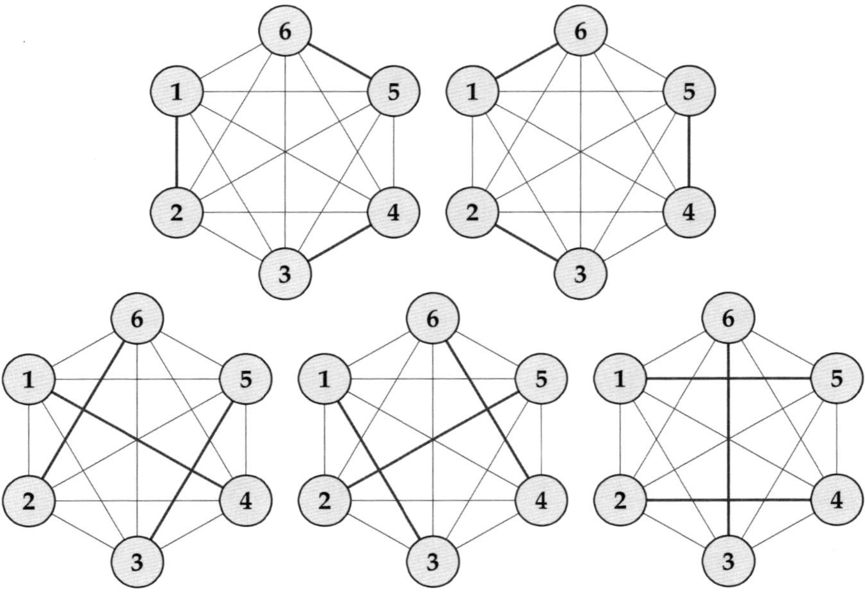

Figure 1.21

This procedure generates five crocodiles which partition the fifteen pairs as shown in figure 1.22.

$$\begin{array}{ccccc}
\{1,2\} & \{6,1\} & \{1,4\} & \{2,5\} & \{3,6\} \\
\{3,4\} & \{2,3\} & \{2,6\} & \{3,1\} & \{4,2\} \\
\{5,6\} & \{4,5\} & \{3,5\} & \{4,6\} & \{5,1\}
\end{array}$$

Figure 1.22

Now let us turn our attention to schoolgirls walking in triples. There is clearly no solution for six schoolgirls, since once we have chosen 123 and 456 as our triples, any other triple will need to repeat a pair. Consider, then, the case when there are exactly nine schoolgirls, walking three abreast in three rows. There are 36 pairs and each crocodile uses up nine of them — is it possible to produce four crocodiles which do not repeat a pair?

Our first crocodile will be formed by taking rows of the table shown in figure 1.23.

1 2 3
4 5 6
7 8 9

Figure 1.23

We can also form a crocodile by taking the columns. Finally we can form two more crocodiles by taking diagonal stripes in the grid, so long as we allow the diagonals to 'wrap around' the grid where necessary. (This is equivalent to drawing the grid on a *torus*.) The results are shown in figure 1.24.

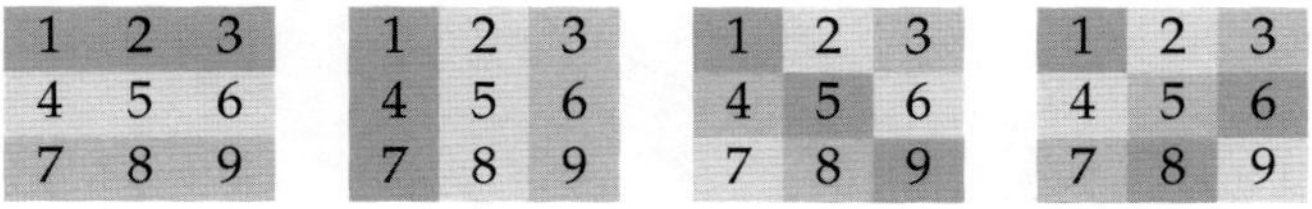

1 2 3	1 2 3	1 2 3	1 2 3
4 5 6	4 5 6	4 5 6	4 5 6
7 8 9	7 8 9	7 8 9	7 8 9

$$\begin{array}{cccc} \{1,2,3\} & \{1,4,7\} & \{1,5,9\} & \{1,6,8\} \\ \{4,5,6\} & \{2,5,8\} & \{2,6,7\} & \{2,4,9\} \\ \{7,8,9\} & \{3,6,9\} & \{3,4,8\} & \{3,5,7\} \end{array}$$

Figure 1.24

We have another geometry. It consists of nine points and twelve lines, with the following properties:

(i) Every point lies on exactly four lines;

(ii) Every line contains exactly three points;

(iii) Two distinct points lie on exactly one line;

(iv) Two distinct lines are either parallel or they pass through exactly one point.

The third and fourth of these properties mirror exactly what we would expect in plane Euclidean geometry. The major difference is that this geometry is finite.

We now introduce another finite geometry based on an equilateral triangle. The points are the vertices of the triangle, the midpoints of the sides and the centroid. The three sides are lines, as are the medians, and there is a seventh 'line' which passes through the three midpoints.

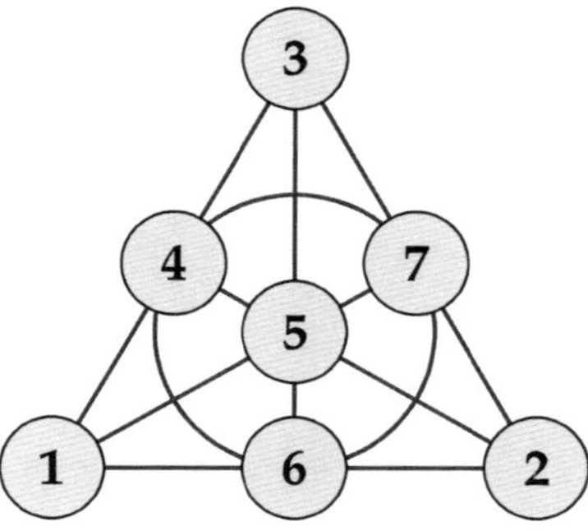

Figure 1.25

This figure is known as the *Fano plane*. It has seven points and seven lines, with the following properties:

(i) Every point lies on exactly three lines;

(ii) Every line contains exactly three points;

(iii) Two distinct points lie on exactly one line;

(iv) Two distinct lines intersect in exactly one point.

This geometry is more symmetrical than the nine point configuration, since every pair of lines intersects in a unique point. It gives rise to a set of seven triples.

$$\{1,2,6\}\ \{2,3,7\}\ \{3,4,1\}\ \{4,5,2\}\ \{5,6,3\}\ \{6,7,4\}\ \{7,1,5\}$$

If we think of the numbers as schoolgirls, we see that every pair is represented once in these triples, and every individual appears three times. It is, however, impossible to reorganise them into crocodiles — this was obvious from the start, since 7 is not a multiple of 3. We then see that

(i) every group contains the same number of girls — three in this case;

(ii) every girl belongs to the same number of groups — again three;

(iii) any pair of girls belongs to exactly one group.

An arrangement like this is known as a *Steiner triple system* (abbreviated to STS). Note that the nine point geometry also gives rise to a STS shown again in figure 1.26.

$$\begin{matrix} \{1,2,3\} & \{1,4,7\} & \{1,5,9\} & \{1,6,8\} \\ \{4,5,6\} & \{2,5,8\} & \{2,6,7\} & \{2,4,9\} \\ \{7,8,9\} & \{3,6,9\} & \{3,4,8\} & \{3,5,7\} \end{matrix}$$

Figure 1.26

If we think of this as twelve separate groups containing three girls, we see that

(i) every group contains three girls;

(ii) every girl belongs to four groups;

(iii) each pair of girls belongs to exactly one group.

However, this arrangement has the additional property that

(iv) the groups can be arranged into four crocodiles, each of which contains all nine girls.

This extra structure is what characterises a *Kirkman triple system*.

1.7 Solving the problem

Problem 1.6 is therefore about constructing a Kirkman triple system with 15 elements. The Fano plane has certainly made a start on this, in that it gives us seven triples involving seven of the schoolgirls. We are, however, aiming for 35 triples, so we need another 28 of them which involve the other eight schoolgirls.

This is reminiscent of the earlier situation where we had a set of eight schoolgirls arranged into crocodiles. They were named from 1 to 8, but it is a trivial matter to add 7 to each name and produce the arrangement shown in figure 1.27. The crocodiles have been reordered in a more natural way.

{■, 8, 15}	{■, 9, 15}	{■, 10, 15}	{■, 11, 15}	{■, 12, 15}	{■, 13, 15}	{■, 14, 15}
{■, 9, 14}	{■, 10, 8}	{■, 11, 9}	{■, 12, 10}	{■, 13, 11}	{■, 14, 12}	{■, 8, 13}
{■, 10, 13}	{■, 11, 14}	{■, 12, 8}	{■, 13, 9}	{■, 14, 10}	{■, 8, 11}	{■, 9, 12}
{■, 11, 12}	{■, 12, 13}	{■, 13, 14}	{■, 14, 8}	{■, 8, 9}	{■, 9, 10}	{■, 10, 11}

Figure 1.27

Of course, these are pairs rather than triples. However, we still have at our disposal the numbers 1 to 7 from the Fano plane and we hope to use these to fill in the grey boxes in the figure. Supplement each pair in the top row by the number corresponding to their crocodile.

What number can be added to the first column of the second row? It cannot be 2, since we already have 2 9, so insert 3 and cycle. For the third row, it cannot be 3 or 4, so insert 5 and cycle. In the fourth row, adding 2 and cycling works. Finally create a new fifth row consisting of the missing Fano numbers in triples — note that this also displays the cyclical behaviour of previous rows. The result is shown in figure 1.28.

{1, 8, 15}	{2, 9, 15}	{3, 10, 15}	{4, 11, 15}	{5, 12, 15}	{6, 13, 15}	{7, 14, 15}
{3, 9, 14}	{4, 10, 8}	{5, 11, 9}	{6, 12, 10}	{7, 13, 11}	{1, 14, 12}	{2, 8, 13}
{5, 10, 13}	{6, 11, 14}	{7, 12, 8}	{1, 13, 9}	{2, 14, 10}	{3, 8, 11}	{4, 9, 12}
{2, 11, 12}	{3, 12, 13}	{4, 13, 14}	{5, 14, 8}	{6, 8, 9}	{7, 9, 10}	{1, 10, 11}
{6, 7, 4}	{7, 1, 5}	{1, 2, 6}	{2, 3, 7}	{3, 4, 1}	{4, 5, 2}	{5, 6, 3}

Figure 1.28

This is the required Kirkman triple system.

Exercise 1d

1. Take the Fano plane as labelled above on page 33 and relabel it by cycling the numbers 1 to 7. What happens?

2. Let F_1 be a Fano plane which is labelled using the numbers 1 to 7. Create a second Fano plane F_2 with points labelled using the names $\{126, 134, 157, 237, 245, 356, 467\}$ with the following property:

   There is a one-to-one correspondence between the labels for F_1 and F_2 such that two points are on the same line in F_1 if, and only if, the corresponding lines meet at the corresponding point in F_2.

3. Three non-collinear points in the Fano plane are said to form a circle, and a line is a tangent to the circle if it passes through exactly one point on it.

   (a) How many circles are there in the Fano plane, and how many through each point?

   (b) How many tangents to a given Fano circle are there at a point on it?

4. Label the Fano plane with ordered triples of binary digits 001, 010, 011, 100, 101, 110, 111, so that, for any two points p and q, the third point on the line containing them is labelled $p + q$, adding pointwise modulo 2 so that, for example, $100 + 101 = 001$.

5. A *finite projective plane* is a finite non-empty set P of points, with a non-empty set of subsets called lines, satisfying the following properties:

   - through any two points there is a unique line;
   - any two lines share a unique point.

   (a) Prove that there is a set of four lines, no three of which pass through the same point.

   (b) Prove that every line contains the same number of points, that every point has the same number of lines through it, and there are the same number of points and lines.

We write this number as $n+1$ and we say that the projective plane has order n. The Fano plane is a finite projective plane of order 2.

(c) Prove that a finite projective plane of order n has n^2+n+1 points and n^2+n+1 lines.

(d) Extend the geometry shown in figure 1.24 on page 32 to form a finite projective plane of order 3.

(e) The game *Dobble* consists of a pack of 55 cards, each of which displays 8 out of a total of 57 different symbols. Any two cards have exactly one symbol in common. The aim of the game is to identify matching symbols, which is not as easy as it sounds, as they are in random orientations and of different sizes.
Interpret this game in terms of a projective plane and comment on the number of cards in the pack.

Chapter 2

Graph Theory

2.1 Königsberg

We will start our discussion of graph theory with a story which has become part of mathematical folklore. In the Prussian town of Königsberg, now Kaliningrad in Russia, two large islands in the Pregel river were connected to each other and to the river banks by a total of seven bridges as shown in figure 2.1.

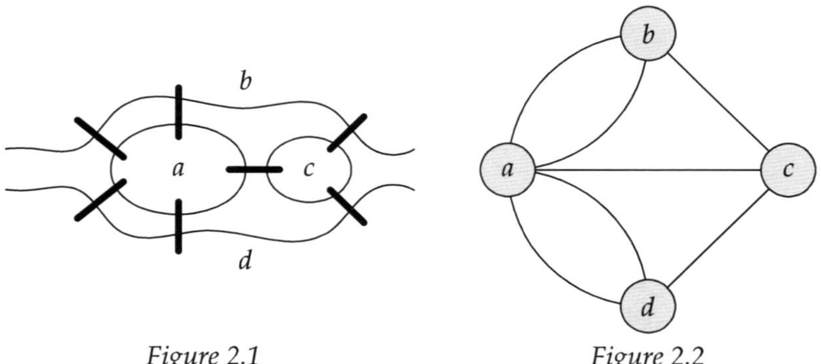

Figure 2.1 *Figure 2.2*

The residents of the town were intrigued by the following question.

Problem 2.1

Is it possible to go for a stroll around Königsberg which crosses each of the seven bridges exactly once?

In 1735 Leonhard Euler proved that the answer is 'No'. None of the residents of Königsberg can ever have completed the challenge, but repeated failure does not prove impossibility. To do that Euler needed an argument that explained why no strategy, even one nobody had tried, could succeed.

The exact location and length of the bridges is irrelevant. All that matters is the connections between the four regions of the city, labelled a, b, c and d in figure 2.1.

The key idea is to focus on the number of connections there are to each area. We note that region a is joined to the rest of the city by five bridges, and that the other three regions boast three bridges each.

Next we imagine going for a stroll around the city which uses no bridge more than once. At the end of the journey we consider each of the four regions in turn, and ask how many of that region's bridges we crossed. For the regions we started and ended in, the answer could be any number between one and five, but there are at least two other regions to consider. Each visit to one of these regions uses two bridges, one on the way in and one on the way out, so for these regions the number of bridges used is an *even* number.

All four regions of Königsberg are blessed with an *odd* number of bridges, so our walk cannot have used all the bridges.

We can replace figure 2.1 with something more abstract in order to make the analysis more transparent. All that matters is which regions are connected to which, so all the relevant information is contained in figure 2.2. The regions have become labelled circles and the bridges have become the lines between them.

Let us consider the simplified maps of the fictional towns Königinsberg and Prinzessinsberg shown in figure 2.3 and figure 2.4 respectively. Again we are interested in whether a journey crossing each bridge exactly once exists.

With Königinsberg we are certainly in luck. If we visit the regions a, b, c, d, e, c, f in that order and then return to a, we use each of the seven bridges once. The fact that our journey starts and ends in the same place means that we could have started anywhere and crossed all the bridges once before returning to our starting point. For a journey like this to exist, it is clearly *necessary* that each region has an even number of connections to other regions.

The seven bridges of Prinzessinsberg require a little more care. The journey b, e, f, a, b, c, d, e crosses every bridge, but the start and end points are different. For a journey like this to exist it is *necessary* that exactly two regions have an odd number of connections to the rest of the map.

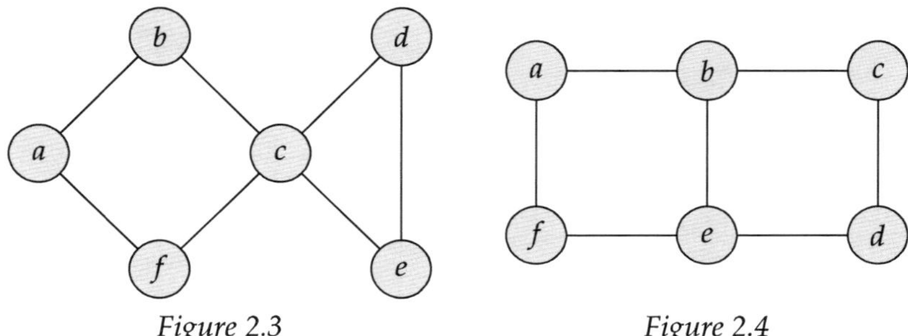

Figure 2.3 *Figure 2.4*

At this point it is natural to ask whether these necessary conditions are also *sufficient*. It seems plausible, but if it is true it will require a careful proof. We will return to this thorny issue in section 2.6.

It is also worth mentioning that diagrams like figure 2.3 need not relate to physical bridges and islands. They could be used to represent any number of scenarios: the way stations are connected on a train network, for example, or which pairs of guests at a small party are acquainted with each other. All we are really studying is the connections between objects, and this is what graph theory is all about.

2.2 Graphs

We start with a definition.

A *graph* is a set of points called *vertices*, each pair of which may or may not be joined by a line called an *edge*. We will use V to refer to the set of vertices, E to refer to the set of edges, and G to refer to the whole graph. The number of vertices a graph has is called its *order*.

The graphs we have just defined have nothing to do with *graphs of functions* plotted on coordinate axes. This might seem confusing, but it is usually clear from the context which meaning is intended. In this chapter the word graph will always refer to a collection of vertices and edges.

The maps of Königinsberg and Prinzessinsberg are both graphs of order six. We might also call them graphs *on* six vertices.

A few comments about the definition of a graph are in order.

- A graph may not have more than one edge between two vertices. (A structure where this is allowed is called a *multigraph*.)
- A graph may not have an edge joining a vertex to itself.
- If an edge joins vertices v and w, we will use the set $\{v, w\}$ to refer to that edge. We will also say that the edge in question *contains* v and w.
- A graph is determined by its vertex set V and its edge set E, so very different diagrams may represent the same graph if they illustrate the same connections.
- Since $\{v, w\} = \{w, v\}$, edges of graphs do not have a specific direction. If we endow each edge with a direction, we obtain a *directed graph* or flow diagram. We shall meet directed graphs again in the context of games in chapter 5.
- If one graph can be turned into another by relabelling the vertices, then the graphs are said to be *isomorphic*. However, we will simply refer to two such graphs as being the *same graph*.

Observant readers will have spotted that the map of Königsberg in figure 2.2 represents a multigraph rather than a graph, since it contains double edges. In this chapter we will restrict our attention to graphs, but the results in section 2.6 can be easily extended to multigraphs.

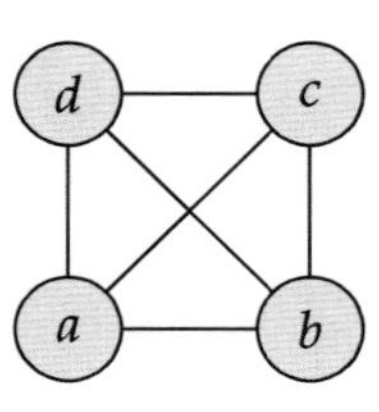

Figure 2.5

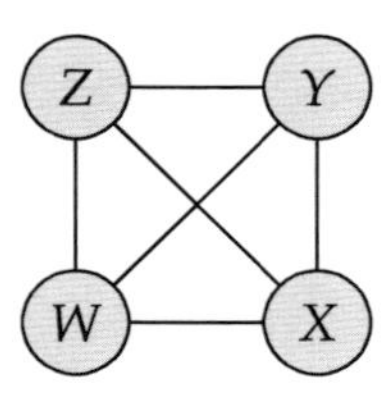

Figure 2.6

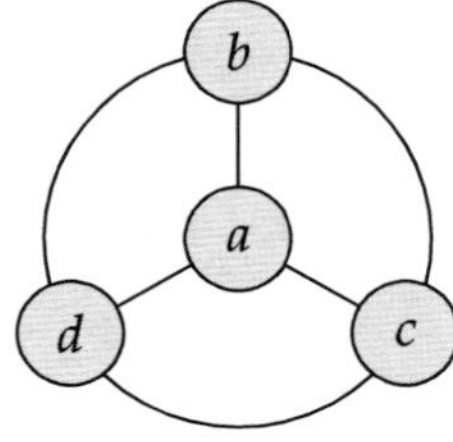

Figure 2.7

Now we turn our attention to figure 2.5, figure 2.6 and figure 2.7. Each one shows four vertices with every possible pair of vertices joined by an edge. Therefore, as far as we are concerned, all three represent the same graph. Note that even though the edges $\{a, c\}$ and $\{b, d\}$ cross in figure 2.5 this does not mean that the graph has another vertex there.

Problem 2.2

Draw six different graphs with order five and five edges.

Since the labels on the vertices are irrelevant, we will represent them by unlabelled dots. The six graphs are shown in figure 2.8. In exercise 2c you will be asked to prove that there are only six graphs on five vertices with five edges. This means the list we have produced is exhaustive, though, of course, diagrams used to represent such graphs might look very different from those shown. The second graph in the list reminds us that the definition of a graph allows it to have vertices which are not joined to any others.

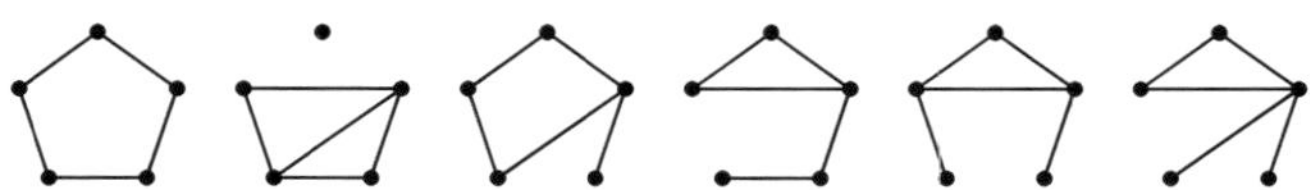

Figure 2.8

2.3 The degree of a vertex

Problem 2.3

Twelve people go to a party, and each person shakes hands with nine others. How many handshakes are there in total?

We represent the party using a graph. The guests are the vertices, and two vertices are joined by an edge if those two guests shake hands. If we count the edges coming from each vertex we get a total of $12 \times 9 = 108$. However, we have counted every edge exactly twice, so the number of handshakes is in fact 54.

This problem is a special case of a result, first proved by Euler, called the *handshaking lemma*. To state the lemma we need another definition.

If v is a vertex in a graph, then the number of edges which contain v is called the *degree* of v, written $\deg(v)$.

Now we are ready for the lemma.

Problem 2.4

Prove that in any graph G with vertex set V and e edges

$$\sum_{v \in V} \deg(v) = 2e$$

.

There are a few symbols to unpack, but this is a straightforward double-counting argument. We will count the number of ends of edges in the graph in two different ways. Counting by the vertices means summing the degrees of all the vertices in the vertex set of G. This is exactly what is denoted on the left hand side of the equation. Counting by the edges we observe that each edge has two ends, so the total number of ends is $2e$.

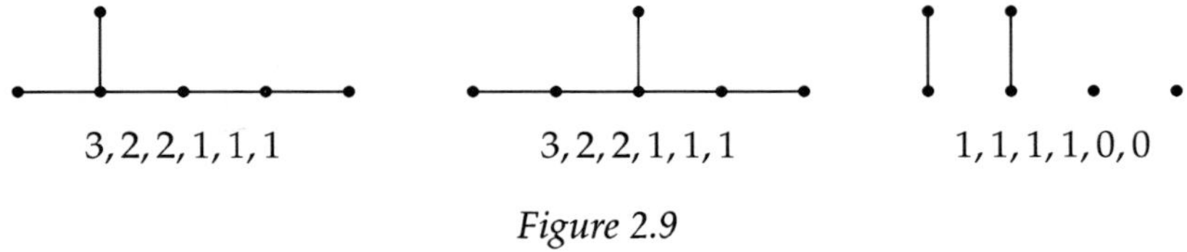

Figure 2.9

If we look again at the Prinzessinsberg graph in figure 2.4, we see that it has two vertices of degree 3 and four of degree 2. Listing the degrees in decreasing order yields the *degree sequence* of the graph, in this case 3, 3, 2, 2, 2, 2. Figure 2.9 shows two different graphs with degree sequence 3, 2, 2, 1, 1, 1 and the unique graph with degree sequence 1, 1, 1, 1, 0, 0.

Exercise 2a

1. What is the maximum number of edges a graph with n vertices can have? (This graph is called the *complete graph* on n vertices, which is written K_n.)

2. There are eleven different graphs on four vertices. Draw them all.

3. Draw all the graphs on five vertices with four edges.

4. Is there a graph with degree sequence 3, 3, 2, 2, 1, 0?

5. Is there a graph with degree sequence 6, 4, 3, 3, 1, 1?

6. Is there a graph with degree sequence 7, 3, 3, 3, 3, 3, 3, 3?

7. Is there a graph with degree sequence 7, 6, 5, 4, 3, 2, 1, 0?

8. Is there a graph with degree sequence 3, 3, 3, 3, 3, 3, 3, 3?

9. There are two graphs with degree sequence 2, 2, 2, 2, 2, 2. Draw them both.

10. Prove that every graph with odd order has an odd number of vertices with even degree.

2.4 Connected graphs, paths and cycles

Figure 2.10 *Figure 2.11*

The graphs shown in figure 2.10 and figure 2.11 both have eight vertices and eight edges. In fact, they have the same degree sequence. However, there is also a striking difference between them. Figure 2.10 seems to be in one piece, while figure 2.11 is made up of two separate pieces. This is clear from the diagrams, but we must remember that a graph is simply a set of vertices and a set of edges. We, therefore, need to find a definition of a 'piece' of a graph, which does not depend on having a drawing of that graph.

We will call the 'pieces' of a graph its *components*. Our intuition suggests that two vertices should be in the same component if it is possible to get from one to the other using edges in the graph. To capture this idea without relying on diagrams, we make the following sequence of definitions.

A *path* is an alternating sequence of distinct edges and vertices which starts and ends with a vertex and where each edge joins the vertices either side of it in the sequence.

Vertices u and v are in the same *component* of a graph if, and only if, there is a path starting at u and ending at v.

If a graph G has only one component then it is called *connected*.

Now we are in a position to distinguish the graphs in figure 2.10 and figure 2.11 since only one of them is connected.

It is worth emphasising that paths use distinct vertices and edges. This means that the journey we used to traverse every bridge of Prinzessinsberg is not a path since it uses vertices b and e twice.

Figure 2.12 *Figure 2.13*

Distinguishing the graphs shown in figure 2.12 and figure 2.13 is more subtle. Both are connected and they have the same degree sequence. However, the graph in figure 2.12 has an edge whose removal would separate the graph into two components. Edges with this property are called *bridges*, and the graph in figure 2.13 has none.

We will need yet another new word to describe the difference between the graphs shown in figure 2.14 and figure 2.15.

Figure 2.14 *Figure 2.15*

If we join the two ends of a path together with an edge, we obtain a *cycle*, and if the cycle consists of n edges and n vertices, then we call it an n-cycle. The graph shown in figure 2.14 has two 3-cycles, a 4-cycle and a 5-cycle, while the graph shown in figure 2.15 has two 3-cycles and two 4-cycles. This allows us to distinguish the graphs even though they each have two components and the same degree sequence. We note that our walk around Königinsberg is not a cycle since it uses vertex c more than once.

2.5 Trees

We now turn our attention to an important family of graphs called trees.

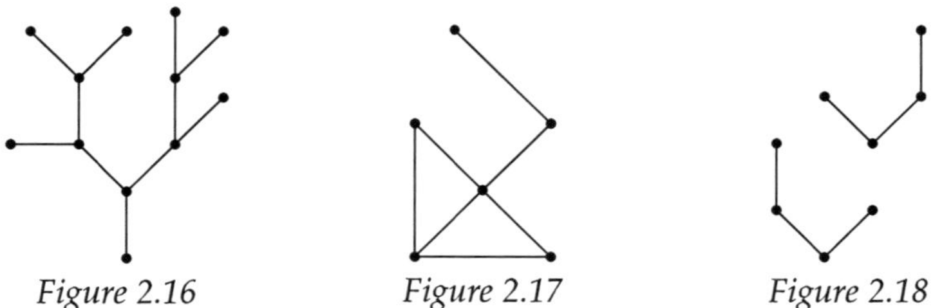

Figure 2.16 *Figure 2.17* *Figure 2.18*

A *tree* is defined as a connected graph which does not contain any cycles.

The graph shown in figure 2.16 is a tree. The graph shown in figure 2.17 fails to be a tree since it contains three cycles, while the graph in figure 2.18 is not connected and hence not a tree.

We call a vertex of degree 1 a *leaf*.

We also say that two vertices connected by an edge are *neighbours*.

Problem 2.5

Prove that every finite tree with at least two vertices has a leaf.

Trying to draw a tree with no leaves seems hopeless. However, we must avoid the logical fallacy that says: 'We tried hard to do X and failed, therefore X is impossible.' Instead, we shall assume that a finite tree with no leaf exists, and then derive a contradiction.

Suppose T is a finite tree with no leaf. T is, by definition, connected, so if we start at any vertex in T we can move to one of its neighbours. Now, having arrived at a vertex via one edge, we can always leave it by another since T has no leaves. However, we can never return to a vertex we have already visited since T, by definition, contains no cycles. Therefore, we can visit as many distinct vertices as we choose, which contradicts the finiteness of T.

A shorter solution to problem 2.5 uses the versatile idea of considering an *extremal* object.

Given a finite tree T, let the vertices $v_1 v_2 \ldots v_l$ form a *longest* path in T. The vertex v_1 cannot be joined to any vertices not in this path else there would be a longer path in T. However, it cannot be joined to any of the vertices in the path, other than v_2, else there would be a cycle in T. Therefore, v_1 is a leaf (as is v_l).

We now turn to a key result about trees.

Problem 2.6

Prove that any tree on n vertices has exactly $n - 1$ edges.

This result might seem too obvious to be worth proving, but the style of argument will be used in more subtle contexts later on.

Like many results in graph theory, this is best proved by induction.

Base

The (only) tree on two vertices has one edge as required.

Step

Assume that every tree on k vertices has $k - 1$ edges.

Now consider an arbitrary tree T on $k + 1$ vertices.

By the result of problem 2.5, T has a leaf which we may call x. If we temporarily remove the vertex x and its edge from T, we obtain a graph on k vertices. Clearly, this graph is still connected and it contains no cycles, so it is a tree. Our inductive hypothesis now shows that it has $k-1$ edges. On returning x and its edge to the graph, we see that T has k edges as required.

It is worth comparing this proof by induction with the first example given on page 235 in the appendix, where it is shown that the sum of the numbers $1, 2, \ldots, n$ is $\frac{1}{2}n(n+1)$.

The logic is, of course, the same, but our tree proof feels a little different. First, what we assume by induction seems much stronger: in the algebraic proof we assumed a result about the single sum $1 + 2 + \ldots k$, while here we assume something about *all* trees on k vertices. The price we pay is that where in one proof we only had to consider the single sum $1 + 2 + \ldots + k + (k+1)$, we now need to consider an *arbitrary* tree on $k+1$ vertices.

This arbitrariness gives rise to another superficial difference in the proofs. It was easy to see how to build the big sum $1 + 2 + \ldots + k + (k+1)$ out of the small sum $1 + 2 + \ldots + k$: we just added the missing term. With the trees, by contrast, it is not so clear how to build the bigger tree from a smaller one, since we do not know which small tree we need to use. This is why we go to the trouble of removing a carefully chosen vertex (a leaf) from the big tree and checking the result was in fact a small tree which we then add the leaf to. This is equivalent to taking the big sum $1 + 2 + \ldots + k + (k+1)$, subtracting $(k+1)$, computing the result and then adding $(k+1)$ back on: absurdly clunky for the sum, indispensable for the trees.

Both of these features are common in proofs by induction in graph theory: we assume something about *all* graphs with k vertices and consider *any* graph with $k+1$ vertices. We then remove a particular vertex from our arbitrary large graph, use our inductive hypothesis, and return the vertex to the graph.

Exercise 2b

1. Prove that there are only six graphs on five vertices with five edges.

2. Construct a connected graph which has a 3-cycle, a 5-cycle and a 7-cycle but no even cycles (that is, cycles of even length).

3. Suppose that a graph G has no cycles, but that adding any edge to G creates a cycle. Prove that G is a tree.

4. Prove that a connected graph is a tree if, and only if, every edge is a bridge.

5. A graph H is a *subgraph* of G if H can be formed from G by removing edges and/or vertices. A subgraph of a graph G which contains every vertex of G and is a tree is called a *spanning tree* of G. Prove that every connected graph has a spanning tree.

6. Prove that if a graph contains a bridge, then it contains a vertex of odd degree.

7. Is it possible for a tree to have the same degree sequence as a graph which is not a tree?

8. What is the maximum number of edges a graph on n vertices can have if it is not connected?

9. A graph G is called *bipartite* if its vertices can be divided into two classes such that no edge joins two vertices in the same class.

   (a) Prove that a bipartite graph cannot contain an odd cycle.

   (b) Prove that a connected graph with no odd cycles is bipartite.

2.6 Eulerian trails and circuits

Let us revisit the bridges in Königsberg, and introduce a few more terms which allow us to express Euler's ideas more succinctly.

We have already met the term *path*, but now we note that it is one of a group of of related definitions.

- A *walk* is an alternating sequence of vertices and edges, starting and ending with vertices, where each edge joins the vertices either side of it in the sequence.
- A *closed walk* is a walk whose start and end vertices are the same.
- A *trail* is a walk whose edges are all different.
- An *Eulerian trail* is a trail which uses every edge in a graph.
- A *circuit* is a trail whose start and end vertices are the same.
- An *Eulerian circuit* is a circuit which uses every edge in the graph.
- A *path* is a trail whose vertices are distinct.
- A *cycle* is a path together with an extra edge joining its first and last vertices.

Graphs with Eulerian circuits are called *Eulerian graphs*, while those with Eulerian trails but not Eulerian circuits are called *semi-Eulerian*.

We will call a vertex with odd degree an *odd vertex* and define *even* vertices similarly.

Euler's reasoning showed that, in an Eulerian trail or circuit, every visit to a vertex which is not the start or end point uses exactly two edges containing that vertex, and thus that all these *transitional* vertices must be even. In particular, an Eulerian graph has no odd vertices, while a semi-Eulerian graph has precisely two odd vertices. A rather more obvious condition is that every Eulerian or semi-Eulerian graph must be connected. Now we return to the question we posed at the start of the chapter and show that these *necessary* conditions are, in fact, *sufficient*. The proof is fairly delicate, and relies on the fact that trails, despite having different edges, may use the same vertex more than once.

Problem 2.7

Prove that if a connected graph G has no odd vertices, then it has an Eulerian circuit.

Since an Eulerian circuit contains all the edges in the graph, it might be worth considering a longest circuit in the graph. We might then try to argue that if this circuit does not use every edge, then it can be extended. Since this would contradict the maximality of the circuit, we could then conclude that the longest circuit is Eulerian.

This can be made to work, but extending circuits is slightly trickier than extending trails, so it is neater to consider a longest trail and prove that this must in fact be an Eulerian circuit.

Let $T_{max} = v_0 e_0 v_1 e_1 \ldots e_{l-1} v_l$ be a longest trail in the graph.

Now we ask how many edges in T_{max} contain v_l. Certainly e_{l-1} does, and there may be more if some of the earlier vertices in the trail equal v_l. Indeed, for each v_i equal to v_l with $1 \leq i \leq l-2$, we have two more edges in T_{max} which contain v_l. Now we ask whether $v_0 = v_l$. If it does not, then an odd number of edges in T_{max} contain v_l. Since v_l is assumed to have even degree, this means the trail can be extended. This contradicts the maximality of T_{max}, and we conclude that $v_0 = v_l$ so T_{max} is a circuit.

Now we suppose, for contradiction, that there is some edge e_* in G which is not in T_{max}. Our aim is to use this edge to extend T_{max}, thus contradicting its maximality. Since G is connected, there are paths from each end of e_* to v_0, so we may form a path with starts by going along e_* and then makes its way to v_0. We follow this path until it first uses a vertex in T_{max} (which may be when in reaches v_0 or may be earlier). We then go round the whole circuit T_{max} to obtain a trail which is longer than T_{max}, giving the desired contradiction.

Problem 2.8

Prove that if a connected graph G has exactly two odd vertices, then it has an Eulerian trail.

It is not too hard to adapt the solution to the previous problem to show that a connected graph with two odd vertices must be semi-Eulerian. However, we can also deduce this fact from the previous result.

Suppose that G is a connected graph and that v and w are the only odd vertices in G. We form a new graph G' by adding a vertex called x and edges $\{v, x\}$ and$\{x, w\}$. The result is a connected graph where every vertex has even degree, so it has an Eulerian circuit. If we walk round this circuit starting and ending at x and then remove x and the first and last edges of the walk, we obtain the desired Eulerian trail.

We note that the extra vertex x was necessary if we wish to restrict our discussion to graphs. We cannot simply add an edge between v and w, since if this edge is already in the graph, then adding it again creates a multigraph. Another option would be to add $\{v, W\}$ if it is not already in G and remove it otherwise. However, if we adopt this approach we must consider the case where $\{u, v\}$ is a bridge separately. This is not hard but could easily be forgotten, leading to a flawed proof. The extra vertex pays its way in terms of elegance.

We have managed to characterise graphs where it is possible to visit each edge exactly once: they are the connected graphs with at most two odd vertices. It now seems reasonable to ask about graphs where we can visit each vertex exactly once.

A cycle which includes every vertex in a graph is called a *Hamiltonian cycle* and the following problem gives a sufficient condition for a graph to have one.

Problem 2.9

Prove that if a graph G has n vertices each of which has degree at least $\frac{n}{2}$, then G has a Hamiltonian cycle.

Our first thought might be to choose a longest cycle in the graph and try to prove that it contains every vertex. However, we can draw on our experience with Eulerian circuits. In that case it was neater to consider a longest trail than a longest circuit, so for this problem it might be wise to consider a longest path, rather than a longest cycle. We will use this path to construct a cycle and show that, if this cycle is not Hamiltonian, then our longest path can be extended, which is absurd.

Let P be a longest path in G, and say that its vertices are $v_0, v_1, \ldots, v_m$.

Our first observation is that this longest path cannot be very short. Indeed, if v_m had a neighbour which was not already listed in the vertices of P, then we could extend P, contradicting its maximality. Since v_m has at least $\frac{n}{2}$ neighbours, this implies that $m \geq \frac{n}{2}$.

Now we describe a way of turning this path into a cycle. Go along the path in the natural order $v_0v_1 \ldots v_i$, until we reach some special vertex v_i. If v_i is a neighbour of v_m, we can then go straight to the end of the path and work our way along it in the opposite order: $v_mv_{m-1} \ldots v_{i+1}$. If v_{i+1} is a neighbour of v_0, we can then complete the cycle. Figure 2.19 shows an example of the procedure with $n = 7$ and $i = 3$. In the top diagram the edges in the path are solid, and two other edges are dotted. In the lower diagram the edges in the cycle are solid.

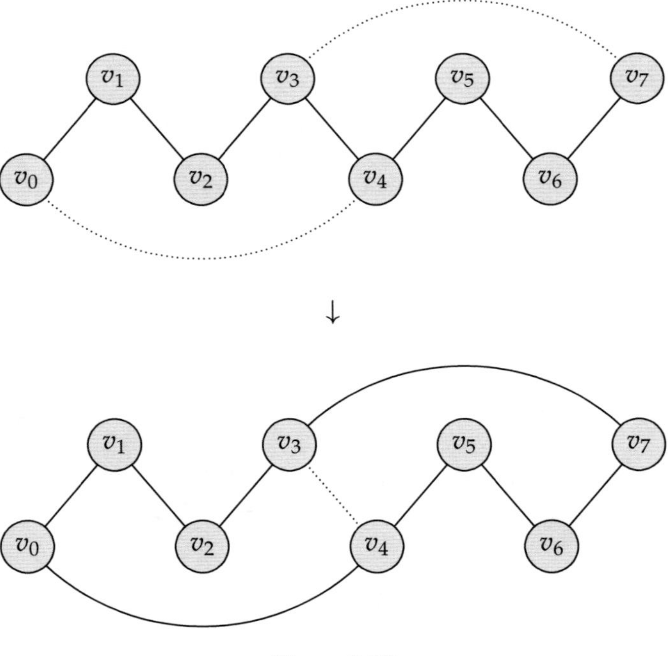

Figure 2.19

Thus, our path can be made into a cycle provided there is some number i such that

(i) $\{v_i, v_m\}$ is an edge of the graph;

(ii) $\{v_{i+1}, v_0\}$ is an edge.

Since $deg(v_m) \geq \frac{n}{2}$, more than half the possible numbers i satisfy (i) and since $deg(v_0) \geq \frac{n}{2}$ more than half the possible i satisfy (ii). This means there must be an i satisfying both criteria, so the path can be made into a cycle.

Our cycle contains $m + 1 > \frac{n}{2}$ vertices, so if there is a vertex w not in the cycle, then it must have a neighbour in the cycle since the number of vertices not in the cycle is less than the degree of w. However, we can now start at w and then go round the cycle to construct a path of length $m + 1$.

This contradicts the definition of P and completes the proof.

Having minimum degree $\frac{n}{2}$ is a neat sufficient condition for a graph to be Hamiltonian, but it is certainly not necessary: a cycle is Hamiltonian and has minimum degree two. We would love to have criteria for Hamiltonicity which were necessary and sufficient, but, in sharp contrast to the situation for Eulerian graphs, no such criteria are currently known. In fact, the existence or not of Hamiltonian cycles in certain families of graphs is an area of current mathematical research.

2.7 Planarity

In 1884 Edwin A. Abbott published a wonderful novel about a two-dimensional world called 'Flatland'. Many of the joys and trials of life in Flatland are vividly described by Abbott, but here we will allow ourselves to consider a problem faced by its inhabitants which he does not address directly.

Imagine that, in a small hamlet in Flatland, there are three houses, unimaginatively labelled A, B and C, and a single source of water, as well as one of gas and one of electricity. The residents would like to connect each house to all three utilities using appropriate pipes and cables. However, they do not want to mix water, gas and electricity, so would like to make the nine connections without the pipes and cables crossing each other. A failed attempt is shown in figure 2.20: the dotted connection cannot be made without crossing an existing line.

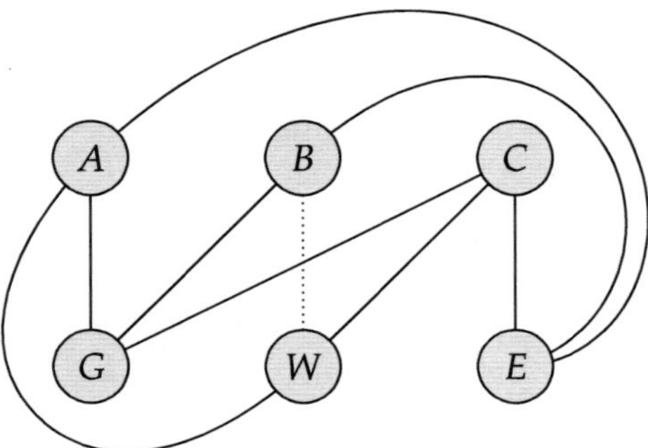

Figure 2.20

In order to state this problem a little more precisely, we will again need to expand our graph-theoretic vocabulary.

We recall that a graph is *bipartite* if its vertices can be split into two subsets such that no edge joins two vertices in the same subset.

If we have a set M of m vertices and a set N of n vertices, and we join every vertex in M to every vertex in N, we obtain a *complete bipartite* graph, written $K_{m,n}$.

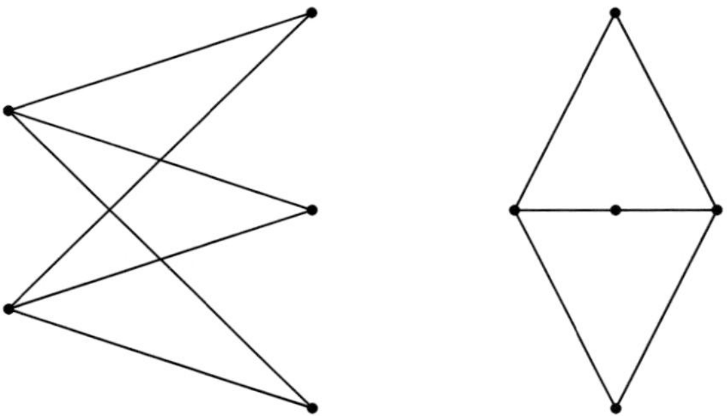

Figure 2.21

Two different drawings of $K_{2,3}$ are shown in figure 2.21. On the left the two vertex classes are visible as two columns, while the drawing on the right has no crossing edges. It bears repeating that these diagrams represent the same set of connections, and thus the same underlying graph.

A graph that is drawn in the plane in such a way that none of its edges cross is called a *plane graph* (or sometimes a *plane drawing* of the graph). A graph which can be drawn as a plane graph is called *planar*.

This definition means that $K_{2,3}$ is a planar graph, even though it is often drawn with intersecting edges. One plane drawing is all that is needed.

The residents of Flatland are trying to draw a complete bipartite graph, since they want to join every house to every utility. We can therefore state their problem very succinctly.

Problem 2.10

Is $K_{3,3}$ planar?

After some experimentation, we abandon our search for a plane drawing of $K_{3,3}$ and start looking for a proof that no such drawing exists. Let us imagine that we have such a drawing. This drawing certainly shows houses A and B each connected to all the utilities. More formally, our supposed drawing of $K_{3,3}$ will contain a drawing of the subgraph $K_{2,3}$. This drawing will look something like figure 2.22 in the sense that it will show six non-crossing edges. This, however, presents us with a problem. The drawing of $K_{2,3}$ divides the plane into three regions, two finite and one infinite, and whichever region house C lies in, it is separated from one of the utilities by the lines that have already been drawn. Therefore, $K_{3,3}$ is not planar.

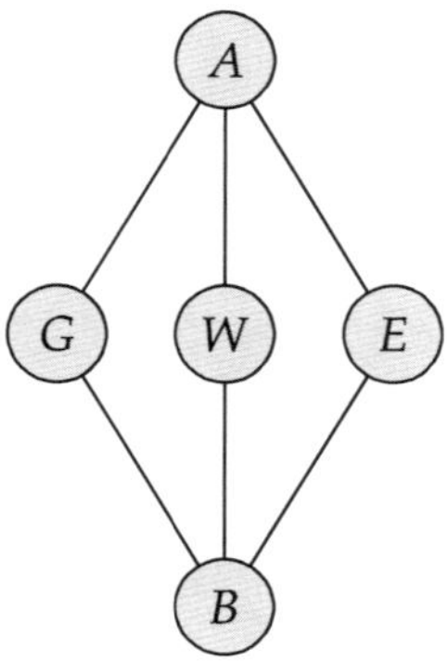

Figure 2.22

The argument we have just given is certainly correct, but it could be criticised for being rather informal. In particular, the claim that every drawing of $K_{2,3}$ 'looks like' figure 2.22 is not terribly precise. What exactly do we mean? Making this proof rigorous by modern standards is possible, but we will not do so here. Later we will provide an alternative proof that $K_{3,3}$ is not planar, which, while still somewhat informal, feels a bit more respectable. To do this we will need to look at the differences between planar and non-planar graphs more generally.

One distinguishing feature of plane graphs is that their edges and vertices divide the plane into a number of regions called *faces*. We count the infinite region 'outside' the graph as a face so that, for example, a graph consisting of a single cycle has two faces and a tree has one.

This means that there are three natural quantities associated with a plane graph: e, the number of edges; v, the number of vertices and f, the number of faces. If we could find some relationship between these three quantities, we might be able to show that certain graphs cannot be planar. With this in mind we find e, v and f for a few plane graphs, such as those in figure 2.23, and collect the results in table 2.1.

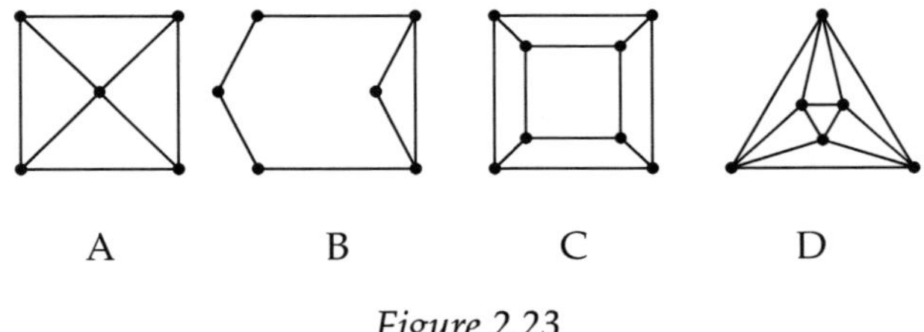

Figure 2.23

This seems to suggest that $e + 2$ always equals $v + f$, but we must be careful: all our examples are *connected* graphs. If we allow graphs that are not connected, then we can add isolated vertices without changing e or f, so any hope of relating e, v and f is lost. Having said that, it may still be that $v - e + f$ is equal to 2 for every connected plane graph. That this is indeed the case was first proved by Euler, and the quantity $v - e + f$ is called the *Euler characteristic* in his honour.

Before we prove that the Euler characteristic is always 2, we should reflect for a moment on how remarkable the result is. For example, there are many millions of different ways to connect 1000 dots with 2000 non-crossing lines, but each creates the same number of regions. More subtly, we can now talk about how many faces a connected planar graph

Graph	e	v	f
A	8	5	5
B	7	6	3
C	12	8	6
D	12	6	8
K_4	6	4	4
$K_{2,3}$	6	5	3
An n-cycle	n	n	2
A tree	$n-1$	n	1

Table 2.1

has. Up until now we have only spoken about the number of faces in a specific plane drawing, and two drawings of the same graph may be rather different. For example, the drawings in figure 2.24 represent the same graph, but only one has a face bounded by a 4-cycle. Euler's result shows that all possible plane drawings of a given planar graph have the same number of faces.

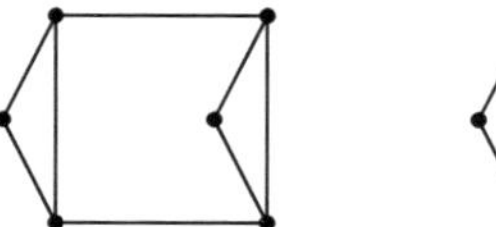

Figure 2.24

Problem 2.11

Prove that for any connected plane graph with e edges, v vertices and f faces, $v - e + f = 2$.

We will use induction on the number of edges in the drawing, and, casting our mind back to our discussion of Eulerian trails, we will take care that our inductive step does not involve removing an edge which might be a bridge.

Base: $(e = 1)$

The only connected graph with one edge has $v = 2, e = 1, f = 1$ so the formula holds.

Step

Suppose that $v - e + f = 2$ for any connected plane graph with k edges.

Consider a connected plane graph G with $k + 1$ edges.

If G contains a cycle, then remove an edge from this cycle.

The result is still a connected plane graph, so for this smaller graph, $v - e + f = 2$ by induction. Returning the edge increases e by 1, increases f by 1 and does not change v. Therefore, the quantity $v - e + f$ is not changed.

If G does not contain a cycle, then it is a tree and thus has a leaf. We remove this leaf and the edge leading to it.

The result is still a connected plane graph, so for this smaller graph, $v - e + f = 2$ by induction. Returning the leaf and its edge increases e by one, increases v by one and does not change f. Therefore, the quantity $v - e + f$ is not changed. This completes the proof.

The Euler characteristic is a fundamental tool for studying planar graphs, and we shall see that it is particularly powerful when combined with a result very similar to Euler's handshaking lemma. The handshaking lemma involved finding the number of edges in a graph by counting the edge ends at each vertex and observing that every edge has two ends. For plane graphs we can perform a similar trick by counting the edges around each face.

Given a plane graph G we will call the set of faces F and, given a specific face $\mathcal{F}$, we will use $|\mathcal{F}|$ to denote the number of edges in the closed walk that is the boundary of $\mathcal{F}$. It is worth noting that $|\mathcal{F}|$ may count the same edge more than once. For example, the graph with two vertices and one edge has a single face $\mathcal{F}$ but $|\mathcal{F}| = 2$. The graph shown in figure 2.25 has three faces and the values of $|\mathcal{F}|$ are 3, 3 and 8.

Figure 2.25

Problem 2.12

Prove that in any plane graph with face set F and e edges

$$\sum_{\mathcal{F} \in F} |\mathcal{F}| = 2e.$$

As with the handshaking lemma, the idea is simple once the symbols are unpacked. If you count the number of edges around every face, you end up counting every edge exactly twice.

Finally we are ready for our second solution to problem 2.10.

We begin by noting that a bipartite graph cannot contain a 3-cycle.

If a plane graph contains no 3-cycles and at least three vertices, then $|\mathcal{F}| \geq 4$ for every face. This means that $\sum_{\mathcal{F} \in F} |\mathcal{F}| \geq 4f$ and so $2e \geq 4f$.

We now have the relations $e \geq 2f$ and $v - e + f = 2$. These imply that $2v + e - 2e \geq 4$ and hence that $2v \geq e + 4$. However, this is false for $K_{3,3}$ since $e = 9, v = 6$. We conclude for the second time that $K_{3,3}$ cannot be planar.

Exercise 2c

1. By mimicking the first proof that $K_{3,3}$ is not planar, find an informal proof that K_5 is not planar. (Recall that K_5 is the graph with five vertices and every possible edge.)

2. Can K_5 be made into a planar graph by deleting one edge?

3. Is $K_{2,4}$ planar?

4. Is K_6 planar?

5. How many edges does $K_{m,n}$ have?

6. Which complete graphs are planar?

7. Which complete bipartite graphs are planar?

8. Prove that, for any planar graph on at least three vertices, $3f \leq 2e$.

9. Prove that, for any planar graph on at least three vertices, $3v - e \geq 6$.

10. Use question 9 to find another proof that K_5 is not planar.

11. What is the maximum number of edges a planar graph on seven vertices can have?

In question 9 of exercise 2c you were asked to prove that, for any planar graph, $3v - e \geq 6$. This is, therefore, a necessary condition for planarity, and it is natural to ask whether this condition is also sufficient. In other words, if a graph satisfies $3v - e \geq 6$ does it follow that it will be planar? Sadly, this turns out to be rather spectacularly false. Indeed, our favourite non-planar graph, $K_{3,3}$ satisfies the relevant inequality. Moreover, if we start with $K_{3,3}$ we can form another non-planar graph by subdividing its edges. In particular, if we take one of the edges of $K_{3,3}$ and divide it into, say, five segments as shown in figure 2.26, we obtain a non-planar graph with $v = 10$, $e = 13$.

The graph in figure 2.27 has the same values of v and e and is clearly planar. This means that we cannot hope for any sufficient planarity condition which only uses the values of v and e.

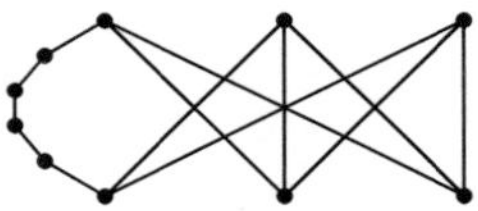

Figure 2.26

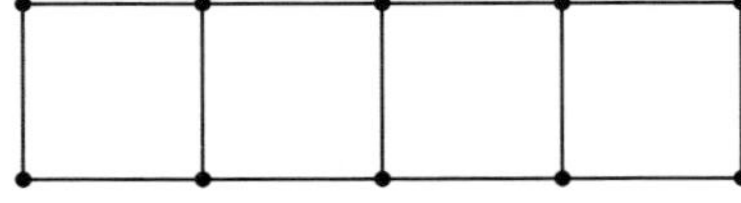

Figure 2.27

Undaunted, we look for another sufficient condition for planarity. Clearly having $K_{3,3}$ or K_5 as a subgraph is an obstacle to being planar, and the example above shows that containing *subdivided* copies of these graphs is also an obstacle. In 1930 the Polish mathematician Kazimierz Kuratowski published a proof that these are, in fact, the only obstacles. Kuratowski's theorem states that a graph is planar if, and only if, it does not contain a (possibly subdivided) copy of K_5 or $K_{3,3}$. This is as good a characterisation of planarity as we could possibly hope for, but the proof is beyond the scope of this book.

2.8 Colouring

We now turn our attention to the celebrated topic of graph colouring. To motivate what follows, imagine a map of a large number of countries. We wish to assign a colour to each of these countries, but want to avoid the potentially confusing situation where two neighbouring countries are assigned the same colour. To be a little more precise we should define a *corner* to be a point where three or more countries meet, and say that countries are only *neighbours* if they share points on their boundaries which are not corners. Thus, for example, we consider the standard colouring of a chessboard to be a legal colouring of a map with sixty-four regions.

Clearly some maps, like the chessboard, can be coloured with only two colours, while others require more. The map shown in figure 2.28 requires four colours, since each of the four countries shares a border with all the others. The natural question to ask at this stage is: 'What is the maximum number of colours which might be required to colour any map?'

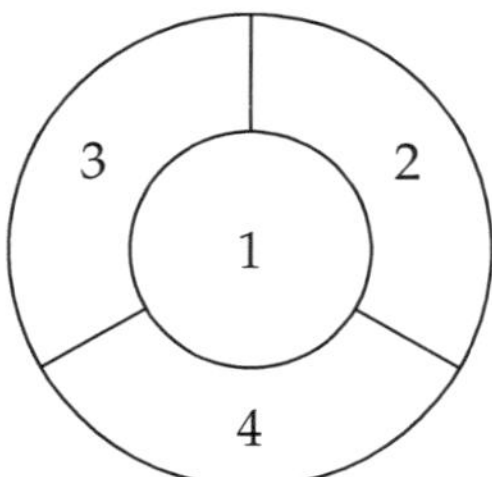

Figure 2.28

The answer is at least four, but there might well be more complex maps that require five, six or many more colours. In 1852 Francis Guthrie asked Augustus De Morgan to provide a proof that no map required more than four colours.

De Morgan failed to provide a proof, and indeed the result is far from straightforward. We will approach it slowly via some weaker related results.

Our first step will be to reframe the question in the language of graph theory. To do this we will imagine that each country on our hypothetical map has a capital city. We also imagine that if two countries share a border then a road between their capitals exists which does not pass through any other countries. If we focus on the capital cities and roads between them, we see that we have constructed a planar graph. This graph is called the *dual* of the map. Since we call a pair of vertices in a graph neighbours if they are joined by an edge, the problem of colouring the original map is reduced to assigning a colour to each vertex of the graph, such that no neighbours are assigned the same colour.

We will call such a colouring with at most k colours a k-colouring.

Now we can succinctly state the so-called *four-colour theorem*.

Every planar graph has a four-colouring.

This result is true, but it is tricky to prove from a standing start. The following problem serves as something of a warm-up.

Problem 2.13

Prove that every planar graph has a six-colouring.

We will use induction on the order of the graph.

Base: $(n \leq 6)$

If a planar graph has at most six vertices, then we can assign each one a different colour to obtain a legal six-colouring.

Step

Suppose every planar graph on k vertices has a six-colouring.

Consider a graph G on $k+1$ vertices.

Our strategy will be to remove a carefully chosen vertex v from G, colour the remaining small graph using our inductive assumption and then return v to the graph and try to assign a colour to v which is different from the colours assigned to all its neighbours.

We observe that this is easy to do if we can choose a vertex v with at most five neighbours, since these neighbours will use at most five colours between them, leaving one for v.

Therefore, it suffices to prove that ever planar graph has a vertex of degree less than six.

Fortunately, this is fairly straightforward: we claim that the average degree in a planar graph is less than six which would not be true if every vertex had degree at least six. All we need is to cast our minds back to question 9 in the previous exercise, where we proved that $3v - e \geq 6$. This means that $6v > 2e$. By the handshaking lemma, $2e$ is the sum of all the degrees in G so the average degree is less than 6 as required. This completes the inductive step and hence the proof.

This is already a big achievement, and with some more work we can go further still.

Problem 2.14

Prove that every planar graph has a five-colouring.

We will try and mimic our proof of the six-colour theorem and proceed by induction on the order of G.

Base: $(n \leq 5)$

If a planar graph has at most five vertices, then we can assign each one a different colour to obtain a legal five-colouring.

Step

Suppose every planar graph on k vertices has a five-colouring.

Consider a graph G on $k+1$ vertices.

We remove a vertex v with degree at most five to form a smaller graph G', five-colour G', and try to return v to the graph.

Now we consider two cases.

Case 1: v's neighbours are assigned at most four distinct colours in the five-colouring of G'.

In this case we can easily return v to the graph and assign it a colour not used by its neighbours.

Case 2: v has five neighbours, each of which is assigned a different colour in the five-colouring of G'.

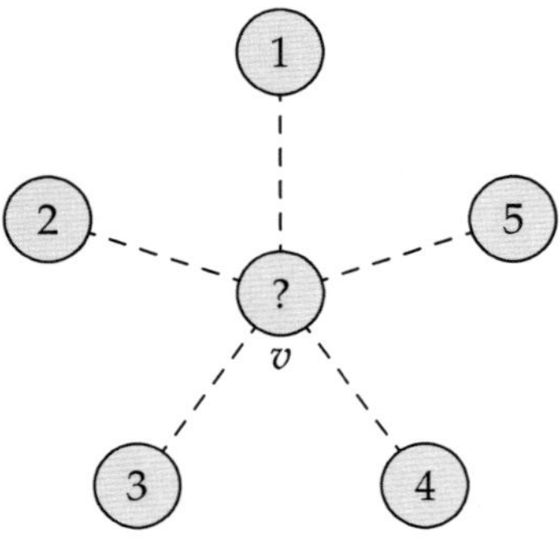

Figure 2.29

If we call our five colours 1, 2, 3, 4 and 5, then this case is shown in figure 2.29. We will use the term *i-neighbour* for a neighbour with colour i. Our aim is to modify the colouring of G' to form a new five-colouring which does not assign all five colours to the neighbours of v. For example, v might have a 1-neighbour, and we might try to recolour this vertex with colour 3, thus leaving colour 1 available for v. Unfortunately, if the vertex we have just recoloured has 3-neighbours, we no longer have a legal five-colouring. One way out is to proceed as follows.

- Start with a 1-neighbour of v, say v', and recolour it with colour 3.
- If the vertex which has just been assigned colour 3 has 3-neighbours, recolour these neighbours with colour 1.
- If any vertices which have just been assigned colour 1 have 1-neighbours, recolour these neighbours with colour 3.
- Repeat this process, exchanging colours 1 and 3 until no further changes are needed.

The effect of this is to swap the colours on any vertex which can be reached from v' via a sequence of vertices whose colours are alternately 1 and 3. This set of vertices is called the *1-3 Kempe chain* containing v' after Alfred Kempe who first introduced the idea. To be sure that the process described above always terminates (for finite graphs), we note that no vertex can be recoloured more than once. A repeated recolouring would require an odd cycle of vertices whose colours are alternately 1 and 3.

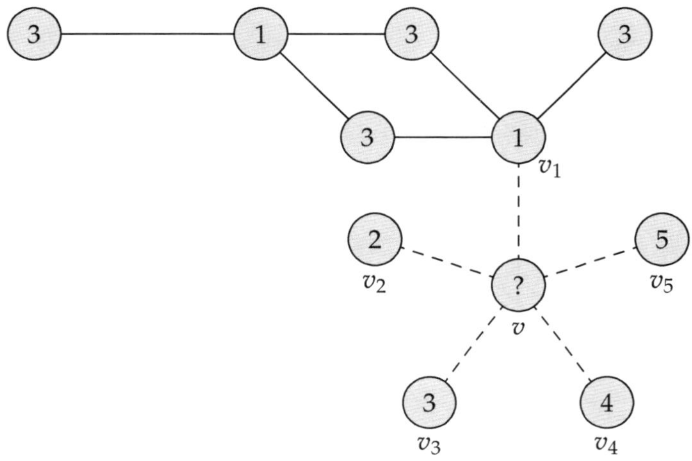

Figure 2.30

However, this cannot exist in the original 5-colouring of G' since then two adjacent vertices would have the same colour.

Kempe chains are useful since, in a legally coloured graph, swapping the colours 1 and 3 on a 1-3 Kempe chain yields another legal colouring.

With this in place we can return to our induction.

We call the neighbours of v, v_1, v_2, v_3, v_4, v_5 in that order, and suppose that they are assigned colours 1, 2, 3, 4 and 5 respectively by the five-colouring of G'. We will try to make colour 1 available for v by swapping the colours on the 1-3 Kempe chain containing v_1.

If v_3 does not belong to this Kempe chain, then colour 1 becomes available for v and we are done. The graph in figure 2.30 has a 1-3 Kempe chain emanating from v_1. In the figure the 1-3 swap has not yet been performed.

Otherwise there is an alternating 1-3 path from v_1 to v_3. In this case, we swap the colours on the 2-4 Kempe chain containing v_2. The crucial observation at this point is that the 1-3 path prevents the 2-4 chain from reaching v_4 as in figure 2.31. This means that colour 2 becomes available for v and the proof is complete.

We have just established the five-colour theorem, and it has cost us a good deal of effort. So before we move on to discuss the four-colour

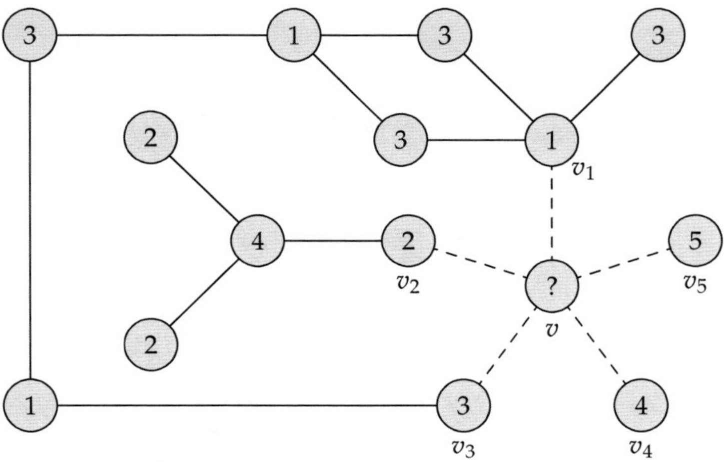

Figure 2.31

theorem, which is harder still, it is worth reviewing all the ideas which fed into the proof.

- Every tree has a leaf (by considering a longest path).
- $\sum \deg(v) = \sum |\mathcal{F}| = 2e$ (by double counting).
- $v - e + f = 2$ (by induction on e).
- $3v - e \geq 6$ (by combining the previous two results).
- Every planar graph has v with $\deg(v) < 6$ (by the result above).
- The six-colour theorem (by induction on v).
- The five-colour theorem (using Kempe chains to modify colourings).

Mathematics is often described as a vast building where new results are constantly added on top of older, more basic ones. We hope that the list above will give the reader some sense of this, without inspiring too much mathematical vertigo.

We now come to one of the crowning jewels of graph theory:

Problem 2.15

Prove that every planar graph has a four-colouring.

As with the five-colour theorem, we proceed by induction on the order of G.

Base: $(n \leq 4)$

If a planar graph has at most four vertices, then we can assign each one a different colour to obtain a legal four-colouring.

Step

Suppose every planar graph on k vertices has a four-colouring.

Consider a graph G on $k+1$ vertices.

We remove a vertex v with degree at most five to form a smaller graph G'. This smaller graph has a four-colouring by induction. Our aim is to orchestrate matters, modifying the colouring if necessary, such that the neighbours of v have at most three different colours.

The four-colouring of G' will already be fit for purpose if it assigns at most three colours to the neighbours of v so we need only consider cases where this does not happen.

If $\deg(v) = 4$ then each colour is used exactly once among the neighbours of v.

If $\deg(v) = 5$ then exactly one colour is used twice among the neighbours of v. We may assume this colour is called colour 1. Since we consider the neighbours of v in order, we may find colour 1 assigned to consecutive or non-consecutive neighbours of v. This means we have three cases to consider.

Case 1a: Vertex v has neighbours v_1, v_2, v_3, v_4 coloured 1, 2, 3, 4 respectively.

Case 1b: Vertex v has neighbours v_1, v_2, v_3, v_4, v_1' coloured 1, 2, 3, 4, 1 respectively.

Case 2: Vertex v has neighbours v_1, v_2, v_3, v_1', v_4 coloured 1, 2, 3, 1, 4 respectively.

In cases 1a and 1b, shown in figure 2.32 and figure 2.33 respectively, we can reuse our Kempe chain argument from the five-colour theorem. In case 1a it is identical. In case 1b we note that if the 1-3 chain containing v_3 does not contain v_1 or v_1' then we can make colour 3 available, while if there is a 1-3 chain from v_3 to either v_1 or v_1', then swapping the colours on the 2-4 chain containing v_2 makes colour 2 available.

Case 2 requires a rather more delicate argument.

If there is no 2-4 path from v_2 to v_4, then a swap on the 2-4 chain containing v_4 makes colour 4 available. So from now on we assume there is such a path. Similarly we may assume that there is a 3-4 path from v_3 to v_4 since otherwise colour 4 could be made available. The situation is shown in figure 2.34.

Now we swap the colours on the 1-3 chain containing v_1. The 2-4 path means that this cannot reach v_3. Finally, swap the colours on the 1-2 chain containing v_1'. The 3-4 path means this cannot reach v_4. Therefore, colour 1 has been made available for v and we are done.

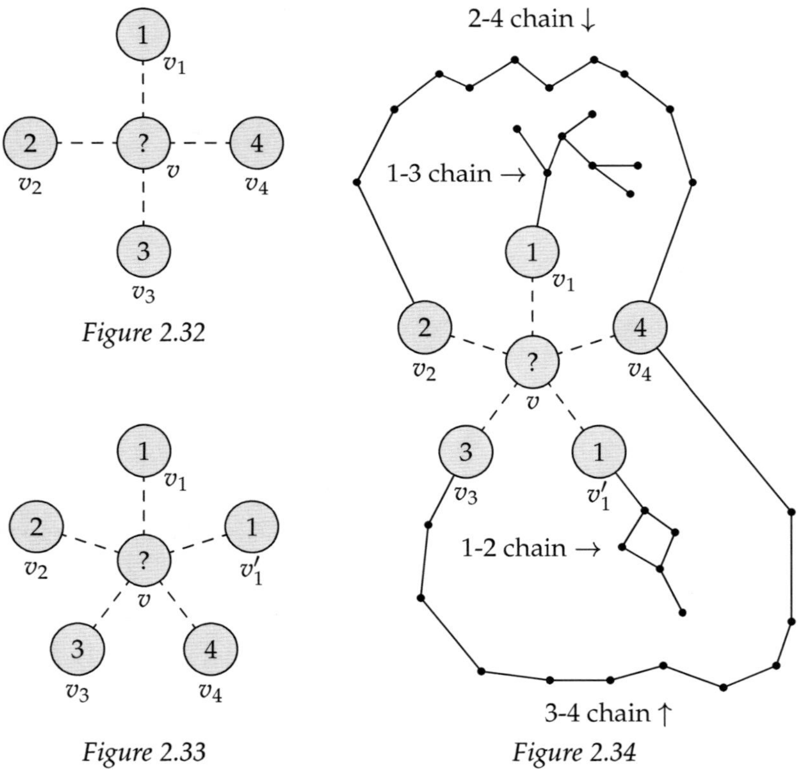

Figure 2.32

Figure 2.33

Figure 2.34

This proof was published by Alfred Kempe in 1879, twenty-seven years after Guthrie first asked De Morgan about the four-colour theorem. This is unsurprising given the intricacy of the argument. What is rather more surprising is that it took a further eleven years for anyone to notice that Kempe's proof was, in fact, wrong!

In 1890 Percy Heawood realised that the proof is fallacious.

Kempe's argument rests on two key facts:

(i) A 2-4 path between v_2 and v_4 prevents a 1-3 path from joining v_1 and v_3. (So v_1 can be changed to colour 3.)

(ii) A 3-4 path between v_3 and v_4 prevents a 1-2 path from joining v_1' and v_2. (So v_1' can be changed to colour 2.)

Each of these claims is true in the original colouring of the graph G. However, Kempe used fact (i) and then performed a 1-3 swap on the 1-3 chain containing v_1. This swap might interfere with the 3-4 path between v_3 and v_4 and might, in some cases, make fact (ii) false.

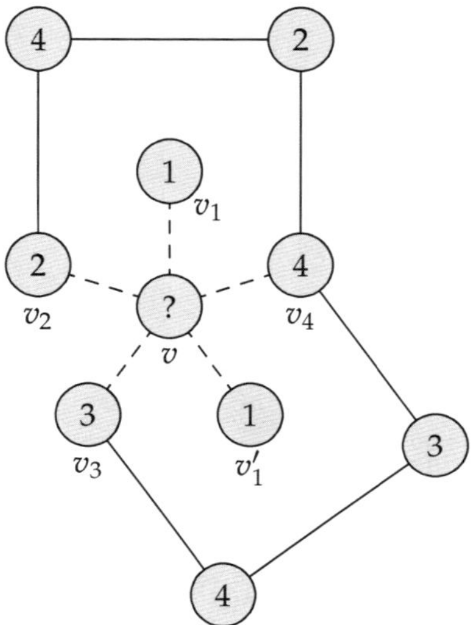

Figure 2.35

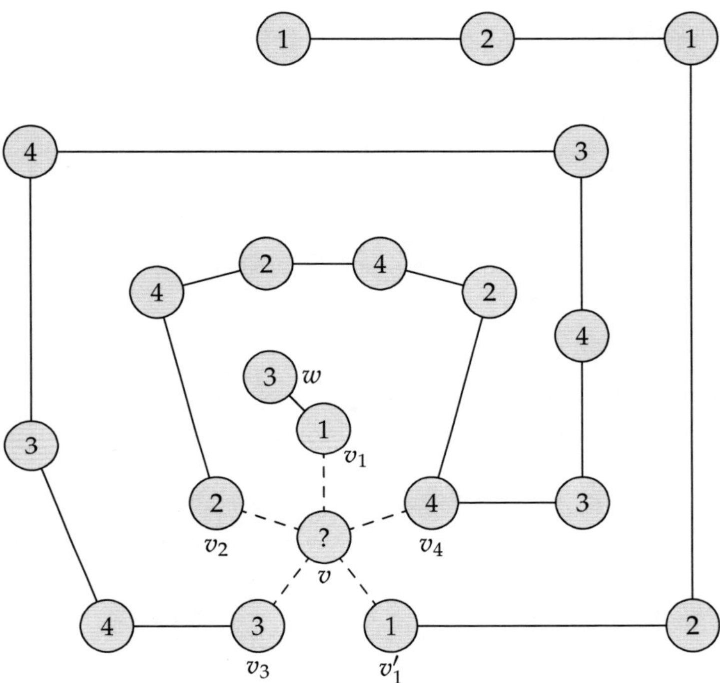

Figure 2.36

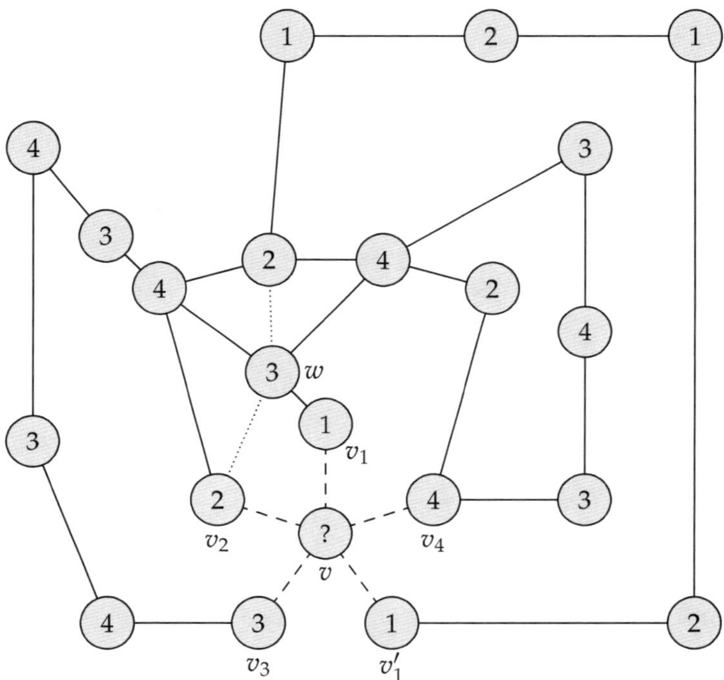

Figure 2.37

If we consider the graph in figure 2.35, the proof look fairly secure. The very short 2-4 and 3-4 chains certainly isolate v_1 from v_3 and v_1' from v_2 as required. We must remember that this figure does not claim to show the whole graph, merely the part needed for the proof. All the vertices other than v might be connected to many other vertices.

The graph in figure 2.36 is slightly more complicated. The 2-4 and 3-4 chains are longer and v_1 is connected to vertex w with colour 3. Moreover, v_1' is part of a substantial 1-2 chain snaking anticlockwise round the outside of the graph. However, things seem to be holding together. The 2-4 chain still protects v_1 from the nasty looking 1-2 tentacle from v_1', while the 3-4 chain, despite going the other way round the graph, still shields v_1' from the tiny 1-3 chain emanating from v_1.

Finally, figure 2.37 shows everything going wrong. The 3-4 path now dips inside the 2-4 path. This in itself is not a disaster, but now the 1-3 swap on the chain containing v_1 makes vertex w colour 1. The dotted lines indicate edges joining w to two vertices with colour 2. Suddenly the 1-2 tentacle from v_1' is able to sneak through and connect with v_2. This means we can no longer ensure that v has no neighbours coloured 1, so no colour can be made available for it. Kempe's proof lies in tatters before us.

Heawood realised that Kempe's methods were sufficient to prove the five-colour theorem, but was unable to fix the proof for four colours. After being considered a theorem for eleven years, the four colour problem was open once more.

In the decades that followed, mathematicians tried and failed to prove the four-colour theorem. There were a number of partial results as well as a number of claimed proofs which turned out to be flawed. It was not until 1976 that a proof by Kenneth Appel and Wolfgang Haken, building on work by Heinrich Heesch, finally emerged.

It is worth making some comparisons between Appel and Haken's proof and Kempe's flawed attempt, though we will not go into any detail. Kempe considered three cases. In each case he argued that a vertex could be removed, that the smaller graph could be coloured, and that the vertex could then be reinserted.

Appel and Haken's proof has a similar logical structure, but where Kempe had three cases, they had 1936. Moreover, each case in Kempe's proof took less than a page to check (albeit incorrectly in one case). By contrast, many of Appel and Haken's cases required thousands of lines of calculations to verify that in that case the graph could be coloured. The task was so vast that they needed to use a computer to carry out much of the checking.

The use of computers in mathematical proofs does not raise many eyebrows these days, but in 1976 it was something rather new. Many people questioned whether a proof that was so long that it could never be checked by a human was even a proof, while others worried that Appel and Haken might have made an error in the hand calculations they needed to build their list of 1936 cases. In any event, it was widely felt that Appel and Haken had shown that the four-colour theorem was true without showing *why* it was true. Over the past thirty-eight years, a number of simpler and more streamlined proofs of the four-colour theorem have emerged, but all of them have made use of computers to do some of the legwork, and at the time of writing no simple pencil and paper proof of this famous result is known.

Exercise 2d

The purpose of this exercise is to give you a chance to prove some theorems about graphs on your own. These will not be used in subsequent chapters, so we recommend that if you find you are stuck, you resist the temptation to look at the solutions too soon, and give yourself time to mull things over. Mathematics is not a sprinting discipline.

1. Every polygon can be decomposed into triangles using non-crossing diagonals. (The proof of this result is the example of complete induction given on page 236 in the appendix.) If we treat both the sides of the polygon and the diagonals used in the decomposition as edges, then we obtain a planar graph which we call a *triangulated polygon*. Prove any triangulated polygon has a 3-colouring.

2. *(The art gallery theorem).* An art gallery is a polygonal room with n walls. The director wishes to position some security guards such that every inch of wall can be seen by at least one of the guards. (Guards can look in all directions, but are not allowed to move.) If the gallery is convex, then a single guard will suffice, but if the floor plan is more complicated then a greater number of guards may be needed. Use the previous question to prove that no matter what shape the gallery is it can always be guarded by $\lfloor \frac{n}{3} \rfloor$ guards.

3. A graph is called *maximal planar* if it ceases to be planar whenever an edge is added to it. Prove that every face of a maximal planar graph (including the outside face) is bounded by a 3-cycle.

4. Prove that a maximal planar graph has at least four vertices with degree less than 6.

5. Prove, by induction on the number of vertices, that every maximal planar graph can be drawn using non-crossing *straight* lines. (*Hint: you will need to use the previous three questions.*)

6. *(Fáry's theorem)* Deduce that every planar graph can be drawn with non-crossing straight lines.

Chapter 3

Serious Counting

Many problems in combinatorics are about counting the number of ways of colouring an object such as a chessboard or a cube. In the case of the chessboard, we might be allowed to colour each of the 64 squares, independently of one another, either black or white. For the cube, the six faces could be coloured red, blue or green. Alternatively, we might be using a wire model representing the twelve edges of a cube, each of which might be coloured in one of four colours. What makes these problems challenging, however, is an extra requirement that rotations (and sometimes reflections) of the patterns are considered as indistinguishable.

3.1 An intuitive approach

We start by considering some problems which can be approached by making a systematic list.

Problem 3.1

Every square of a 2 by 2 grid is coloured either grey or white. How many different outcomes are there?

This is easy. We have two choices four times, so there are sixteen outcomes, as in the following table.

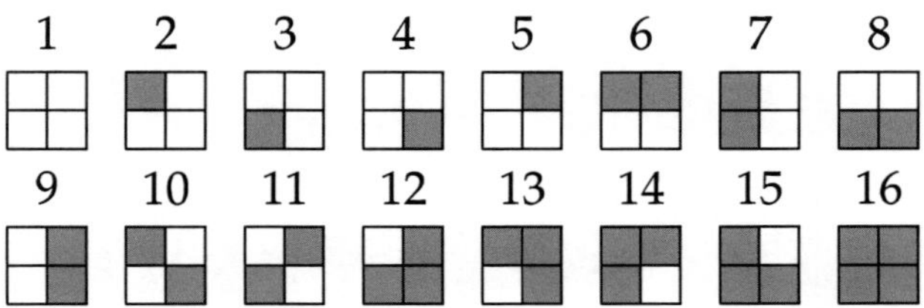

Figure 3.1

Problem 3.2

Every square of a 2 by 2 wooden board, which can be rotated, is coloured either grey or white. How many different outcomes are there?

The important difference between this problem and the last is that rotations of a colouring count as the same. That means that the grids 2, 3, 4 and 5 only count as a single outcome. The same happens with the following sets of grids $\{6,7,8,9\}$, $\{10,11\}$ and $\{12,13,14,15\}$. Finally, grids 1 and 16 cannot be rotated into any of the others, so these are two more outcomes giving a total of six.

Problem 3.3

Every square of a 2 by 2 glass board, which can be rotated and also looked at from either side, is coloured either grey or white. How many different outcomes are there?

Turning the board over and viewing it from the other side is equivalent to reflection. So in this problem not only do rotations of a colouring count the same, but reflections do as well. For example, grids 3 and 4 are reflections of one another in a 'vertical' axis, grids 6 and 8 in a 'horizontal' axis and grids 12 and 14 in a diagonal axis. In the case of a 2 by 2 grid, this makes no difference at all; the outcomes which counted as the same under reflection were already indistinguishable under rotation.

Suppose that we are allowed to use three colours rather than just two.

Problem 3.4

In how many ways can a 2 by 2 square be coloured using three colours, if rotations and reflections are counted as identical?

The first observation is that, unlike the situation in problem 3.3, reflections can be distinguished from rotations.

The two outcomes shown cannot be rotated into one another but can be reflected to one another in a diagonal axis.

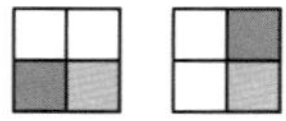

Figure 3.2

There are now 3^4 or 81 colourings, if we ignore the effect of rotations and reflections. Three of them involve only one colour and these are different outcomes. If they involve exactly two colours — say white and dark grey — then, as in problem 3.1, there are 14 of them which reduce to 4 outcomes when we identify rotations and reflections. This takes care of another 42 of the 81 colourings, yielding another 12 outcomes.

The other colourings use two squares of one colour and one square of each of the other two. If, for example, there are two white squares, then there are $\binom{4}{2} = 6$ ways of choosing where the white squares go and then two ways to complete each colouring. These are shown in figure 3.3.

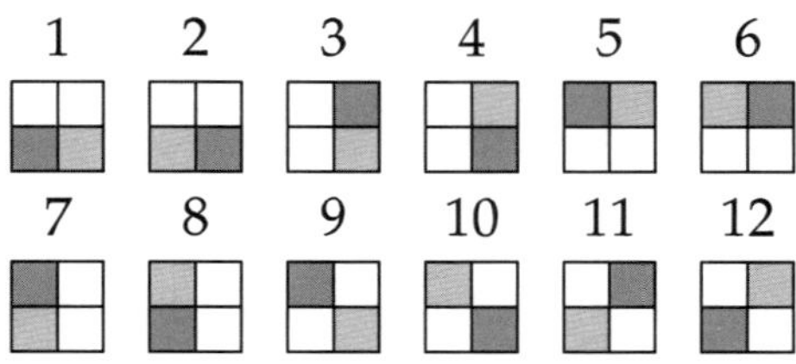

Figure 3.3

We now count how many different outcomes there are when reflections and rotations are counted as identical. The effect is quite drastic, since outcomes 1 to 8 become indistinguishable as do 9, 10, 11 and 12. So there are essentially only two different outcomes with three colours, two of which are white, and when we allow for the rest, we see that there are exactly 6 outcomes.

Hence the total number of outcomes is $3 + 12 + 6 = 21$.

Despite the large number of possible grids, it was possible to solve this problem by splitting into cases and seeing the different effect of the rotations and reflections on these. However, such a hands-on approach becomes intractable with bigger boards. For a 3 by 3 board and 5 colours, there are $5^9 = 1\,953\,125$ grid configurations. In order to deal with colouring problems of this type, we need to develop a theoretical structure, and that is the task of this chapter.

3.2 Equivalence classes

First we identify the set of things which are to be coloured. For example, this might be

- the set of vertices in a square lattice;
- the set of cells in a square grid;
- the set of vertices or sides of a regular polygon;
- the set of vertices, edges or faces of a regular polyhedron.

Since the overall shapes chosen are regular, it is possible to rotate them or reflect them and make 'different' colourings equivalent. This is the difference between problems 3.1 and 3.2.

For the purposes of clarification, we shall use the term *raw colouring* for the situation in problem 3.1. In problem 3.2, different raw colourings are identified if they can be transformed into one another using a rotation. We call such a subset of the set of raw colourings an *equivalence class*. In problem 3.2, we count equivalence classes under rotation, and in problem 3.3, we list the equivalence classes under rotations and reflections. Both these problems concern two colours, but in problem 3.4 we allow three; in general, if there are k colours, we will use the term k-colouring. When we ask for the number of k-colourings of a certain configuration under a set of symmetries, we require the number of different equivalence classes.

Problem 3.5

(a) Find the number of 2-colourings of a 4 by 4 square grid under rotations.

(b) Find the number of 2-colourings of this grid under rotations and reflections.

Let us see what happens with a 4 by 4 square grid whose cells are coloured either black or white. There are $2^{16} = 65\,536$ raw 2-colourings, so there is obviously no possibility of listing them all. We now examine what happens when different raw 2-colourings are counted as equivalent under various conditions.

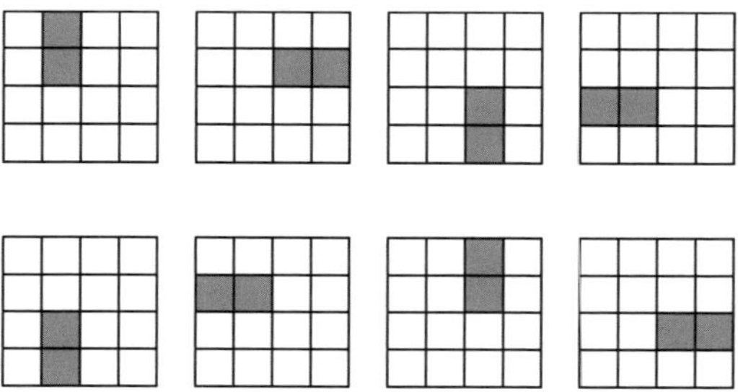

Figure 3.4

In part (a), outcomes which can be rotated into each other will count the same. The first four configurations of the eight above form one equivalence class and the last four form another class. In part (b), these will all count as the same outcome since they can all be transformed to the first under rotation and reflection. We have only one equivalence class containing all eight outcomes.

On the basis of this discussion, it might seem that the number of 2-colourings of a 4 by 4 grid under rotations and reflections is $2^{16} \div 8 = 2^{13}$. Unfortunately this argument is badly flawed, since not all the equivalence classes have eight elements.

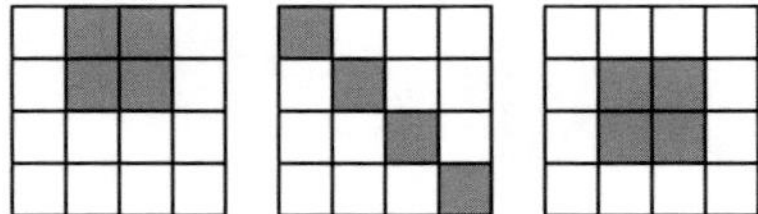

Figure 3.5

The left-hand raw colouring is in a class of four, the middle one is in a class of two and the right-hand one is in a class on its own.

It is this fact — that not all equivalence classes have the same size — which makes colourings hard to count. This is clearly a difficult problem and we postpone the solution until section 3.5.

Exercise 3a

1. The vertices of an equilateral triangle are coloured red, blue or green. Show that there are 11 equivalence classes under rotation, but only 10 if reflections are also allowed. How would these answers change if the sides were coloured instead of the vertices?

2. The vertices of a square are coloured red, blue or green. Show that there are 24 equivalence classes under rotation and 21 under rotation and reflection. How would these answers change if, instead, the sides were coloured?

3. The sides of a non-square rectangle are coloured red or blue. How many equivalence classes are there under rotation alone, and under both rotation and reflection?

3.3 Rotations and reflections

In order to count equivalence classes, we first need to find some way of dealing mathematically with rotations and reflections. We can discuss this in the context of a regular hexagon whose vertices are labelled A to F as shown in figure 3.6.

A hexagon can be rotated into itself by means of an anticlockwise rotation of 60° about its centre. Note that this takes A to B, B to C and so on. If this is repeated, the result is a rotation of 120° which takes A to C, and there are further rotations through 180°, 240° and 300°. After six rotations of 60°, all the vertices will be back to their starting position. This is called the *identity* transformation.

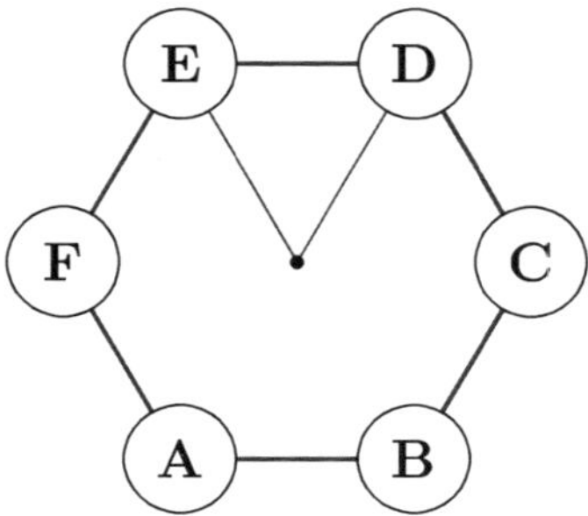

Figure 3.6

If we represent a rotation of 60° by R, we can write the other rotations as R^2, R^3, R^4 and R^5, and it is also sensible to write $R^6 = I$, where I is the identity transformation. So we are beginning to develop an *algebra* of transformations.

However, a hexagon also has six axes of symmetry. Three of them join opposite vertices and the other three midpoints of opposite sides. They are labelled from 1 to 6 in figure 3.7. A reflection in 6 will switch the pairs (C, E) and (B, F) while leaving A and D fixed. A reflection in 1 switches all three pairs $(A, B), (C, F)$ and (D, E).

Suppose we denote by T_r the reflection in axis r. If T_6 is followed by R, the result is T_1, since C goes to E and then to F, and so on. If, however, R is followed by T_6, the result is T_5. Combination of transformations is not commutative.

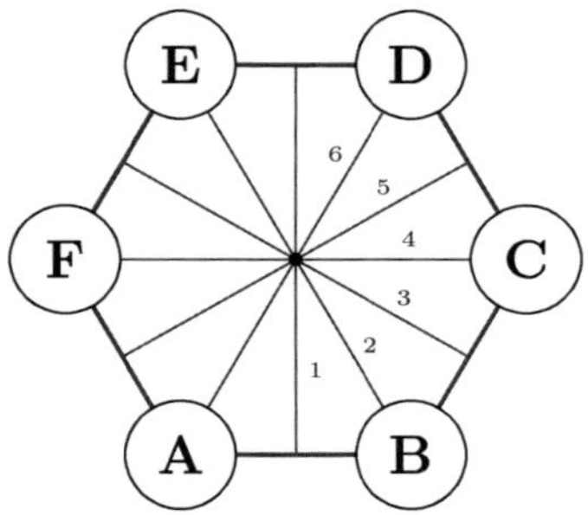

Figure 3.7

This means that we have to be careful in our notation. We shall use the convention that a transformation XY means Y followed by X, which seems counterintuitive, but it is consistent with the way we write functions, where $fg(x)$ means that we do g before f. So we have $RT_6 = T_1$ and $T_6R = T_5$.

If we carry out this calculation for every pair of transformations, we obtain table 3.1. The result of the combination XY will be found in row X and column Y. (The reader should be assured that they will not need to draw such tables in order to use the techniques in this chapter, but it is worthwhile seeing an example.)

We note that $(XY)Z = X(YZ)$ — the combination is *associative*. This is a property of transformations which is proved in the appendix. We also see that the identity element I has the property that $IX = XI = X$ for every X. This is obvious since the identity does not change anything. Finally, for every X there is a unique Y such that $XY = YX = I$. This is known as the *inverse* of X and is written X^{-1}. It is easy to see that each reflection is its own inverse and the inverse of a rotation of 60° is a rotation of 300°, and so on. As already mentioned, XY is not necessarily the same as YX.

Sets with these properties are very important in abstract algebra, and are known as *groups*. The number of elements in a group is called its *order*, and the notation $|G|$ will be used for this, in line with set notation. So the set of rotations and reflections of a hexagon forms a group of order 12, which is known as the *dihedral group* D_6 (although some authors call this D_{12}). It is also true that the set of rotations alone forms a group of order 6, which is known as the cyclic group C_6. We do not propose to do more

	I	R	R^2	R^3	R^4	R^5	T_1	T_2	T_3	T_4	T_5	T_6
I	I	R	R^2	R^3	R^4	R^5	T_1	T_2	T_3	T_4	T_5	T_6
R	R	R^2	R^3	R^4	R^5	I	T_2	T_3	T_4	T_5	T_6	T_1
R^2	R^2	R^3	R^4	R^5	I	R	T_3	T_4	T_5	T_6	T_1	T_2
R^3	R^3	R^4	R^5	I	R	R^2	T_4	T_5	T_6	T_1	T_2	T_3
R^4	R^4	R^5	I	R	R^2	R^3	T_5	T_6	T_1	T_2	T_3	T_4
R^5	R^5	I	R	R^2	R^3	R^4	T_6	T_1	T_2	T_3	T_4	T_5
T_1	T_1	T_6	T_5	T_4	T_3	T_2	I	R^5	R^4	R^3	R^2	R
T_2	T_2	T_1	T_6	T_5	T_4	T_3	R	I	R^5	R^4	R^3	R^2
T_3	T_3	T_2	T_1	T_6	T_5	T_4	R^2	R	I	R^5	R^4	R^3
T_4	T_4	T_3	T_2	T_1	T_6	T_5	R^3	R^2	R^2	I	R^5	R^4
T_5	T_5	T_4	T_3	T_2	T_1	T_6	R^4	R^3	R^3	R	I	R^5
T_6	T_6	T_5	T_4	T_3	T_2	T_1	R^5	R^4	R^3	R^2	R	I

Table 3.1

than the minimum amount of group theory in this book, and we will not be worrying very much about the composition tables.

Exercise 3b

1. List the elements of the group of rotations and reflections of an equilateral triangle.

2. List the elements of the group of rotations and reflections of a square.

3. List the elements of the group of rotations of a regular pentagon.

4. List the elements of the group of rotations of a regular tetrahedron.

3.4 Colourings and transformations

Let us now revisit problem 3.3 in the light of the previous discussion. We have to determine the number of equivalence classes of raw colourings, where two colourings belong to the same class if there is a transformation in the group D_4 which takes one to the other. The sixteen raw colourings are listed again in figure 3.8.

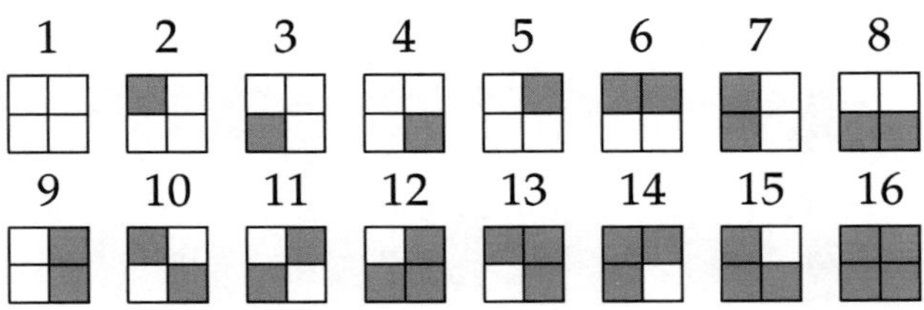

Figure 3.8

We now organise these colourings in a table, whose rows correspond to the eight transformations in D_4. This is done systematically as follows:

- begin by placing the first raw colouring in row I and the first column;
- now identify the raw colouring which arises when this is subjected to each of the rotations or reflections and write its number in the appropriate cell;
- find the next unlisted raw colouring and place this in row I and the next column;
- continue this process until all raw colourings have been listed, and then stop.

The result is shown in table 3.2. Unsurprisingly, colouring 1 appears in every cell in the first column. The second column is more interesting, however, and it turns out that the first unlisted colouring is 6, which forms the first entry in the third column. Eventually every cell in the table is populated.

I	1	2	6	10	12	16
R	1	3	7	11	13	16
R^2	1	4	8	10	14	16
R^3	1	5	9	11	15	16
X	1	3	8	11	13	16
Y	1	5	6	11	15	16
U	1	2	7	10	12	16
V	1	4	9	10	14	16

Table 3.2

If we listed the sixteen raw colourings in a different order, then the top row might change. However, the columns would have the same entries, in a different order, and the table we produce would be effectively the same.

Note that colourings appear several times in the table. For example, the colouring 1 appears eight times, colouring 2 appears twice and colouring 10 appears four times. Colourings which are highly symmetrical have many duplicates.

We now count the 48 cells in the table in two different ways. One way is simply to multiply the number of rows by the number of columns. The other is to add up the number of times each raw colouring appears in the table. Working in numerical order, we obtain the fact that

$$8+2+2+2+2+2+2+2+2+4+4+2+2+2+2+8 = 6 \times 8$$

or, in general terms

$$\sum_r \{\text{Number of appearances of raw colouring } r\} = N \times |G| \qquad (3.1)$$

where we are summing over all raw colourings. Here N is the number of equivalence classes and G is the group of transformations. We know what the order of G is, so the problem of calculating N reduces to that of calculating how many times each raw colouring appears in the table.

We will need to do some work on transformation and colourings, and it is worthwhile introducing a piece of notation. If A is a transformation and i is a colouring, we denote by $A(i)$ the colouring to which i is mapped under A. For example, $R^3(6) = 9$, as you will see from the table.

If we examine the way duplicates turn up in the table for problem 3.3, we see that they never appear in different columns. Can we prove that this always happens? It turns out that this is not too difficult, but it relies on the fact that the transformations form a group. Suppose that we have the same raw colouring appearing somewhere in two different columns headed by i and j. This means that there are two elements of G, which we will call A and B, so that $A(i) = B(j)$. We deduce that $i = A^{-1}B(j)$. However, $A^{-1}B$ is another transformation in the group, so i appears somewhere in column j. However, we know that cannot happen, by the way that we constructed the table.

Now let us suppose that there are duplicate colourings i which appear in the same column of the table. If one of these is the column header, the duplicates will appear in the rows labelled by transformations A for which $A(i) = i$. Including the identity transformation in the count, we see that the number of duplicates in such a row is equal to the number of transformations in G which fix i, which is another way of saying that $A(i) = i$. For example, we have $I(10) = R^2(10) = U(10) = V(10) = 10$ in the column headed by 10. There are four occurrences of the raw colouring 10 and there are four transformations which fix this colouring.

What happens if there are duplicate colourings i in a column headed by a different colouring k? We could choose one of these — that nearest the top, in the row labelled A — and we see that $A(k) = i$. Another duplicate in row B satisfies $B(k) = i$, and it follows by group algebra that $i = BA^{-1}(i) = C(i)$. Each different duplicate will correspond to a different C of this form with $i = C(i)$. Conversely, for each transformation D which fixes i (including the identity) we can form a transformation $E = DB$ such that $E(k) = i$. Hence we see that the number of duplicates in a column is the number of transformations which fix i.

An example of this in table 3.2 is the four 11s in column 10. The first 11 is $R(10)$ and the fourth is $Y(10)$. We can go straight from the first to the fourth by means of the transformation YR^{-1} which is equal to U, reflection in the leading diagonal, and it will be seen that raw colouring 11 is indeed fixed by that transformation.

So the number of appearances of a raw colouring in the table is equal to the number of transformations which fix it.

Equation (3.1) is essentially about counting the number of entries in a table with rows and columns. We can think of it as a table of ordered pairs of the form (raw colouring, transformation which fixes it). Our first idea is to count the number of ordered pairs colouring by colouring, so we will have to count some colourings many times.

At this point, we have a clever idea. Instead of counting colouring by colouring, we can count by *transformations*. Again, we will need to count some transformations more than once, and what will determine this is the number of colourings which are fixed by a particular transformation. We shall call this the *invariant set* for a particular transformation. Now equation (3.1) can be modified to read

$$\sum_{r} \{\text{Number of colourings in the invariant set of } \pi\} = N \times |G|. \quad (3.2)$$

Let us illustrate how this works for table 3.2. Counting by colourings gives table 3.3, while counting by transformations yields table 3.4.

Colouring	Transformations which fix it
1	8
2	2
3	2
4	2
5	2
6	2
7	2
8	2
9	2
10	4
11	4
12	2
13	2
14	2
15	2
16	8

Table 3.3

Transformation	Invariant colourings
I	16
R	2
R^2	4
R^3	2
X	4
Y	4
U	8
V	8

Table 3.4

It is worthwhile introducing some notation. We will use Ω for the set of all raw colourings of some configuration, and Ω_π to mean the invariant set corresponding to a transformation π in the group G. In the example above, $\Omega = \{1,2,3,\ldots,16\}$ and $\Omega_U = \{1,2,4,10,11,12,15,16\}$. Now the double-counting method shows us that $N \times |G| = \sum_{\pi \in G} |\Omega_\pi|$, which can be rearranged to give

$$N = \frac{1}{|G|} \sum_{\pi \in G} |\Omega_\pi|. \tag{3.3}$$

This can be paraphrased in a memorable way:

Burnside's lemma

The number of equivalence classes of colourings for a configuration under a group of transformations is the average size of an invariant set.

This result was proved by the English group theorist William Burnside, but, as he acknowledged, it was discovered some ten years earlier by the German mathematician Ferdinand Frobenius. Misattribution of results is very common in mathematics, and we will use the traditional name. The importance of this lemma is that if we want to count the number of colourings of a configuration under a group of symmetries, all we need to know is the order of the group and the sizes of the invariant sets for the different transformations.

3.5 Using Burnside's lemma

We start by revisiting some well-known problems.

First we find the number of arrangements of n people around a round table. We can think of the seats as vertices of an n-gon which is to be coloured using n distinct colours which correspond to the diners. There are $n!$ raw colourings, and we will identify them under rotation using the cyclic group C_n, which is of order n. The invariant set for I is all the colourings and so $|\Omega_I| = n!$. All the other invariant sets are empty. Hence the number of different colourings is $\frac{n!}{n} = (n-1)!$.

Now we find the number of arrangements of n distinguishable beads on a necklace. This is the same situation except that we are allowing reflections as well as rotations and thus using the group D_n which has $2n$ elements. There is still only one non-empty invariant set, so the number of colourings is $\frac{n!}{2n} = \frac{(n-1)!}{2}$.

Now we revisit question 1 of exercise 3a, which was about 3-colourings of an equilateral triangle. If we stick to rotations, the group is C_3. There are 27 raw colourings.

We have $|\Omega_I| = 27$ and $|\Omega_R| = |\Omega_{R^2}| = 3$ since the only invariant sets are monochromatic. Burnside's lemma gives the number of colourings as $\frac{27+3+3}{3} = 11$.

If we allow reflections as well as rotations, and use the group D_3, we also have $|\Omega_A| = |\Omega_B| = |\Omega_C| = 9$. This is because a colouring which is invariant under reflection can have any colour for the vertex on the median and any colour for the remaining two vertices. By Burnside's lemma, the number of colourings is $\frac{27+3+3+9+9+9}{6} = 10$.

Finally, consider again problem 3.4, which is about 3-colourings of a 2×2 square under D_4. There are 81 raw colourings which we must apportion into invariant sets. It is easy to see that $\Omega_I = 81$ and that $|\Omega_R| = |\Omega_{R^3}| = 3$. For R^2 opposite squares must have the same colour so $|\Omega_{R^2}| = 9$. For reflections in the x or y axes, the upper and lower row or column must be identical, so $|\Omega_X| = |\Omega_Y| = 9$. For the diagonal axes it is a little more subtle. There is a free choice for the two squares on the axis, and the other two must have the same colour, so $|\Omega_U| = |\Omega_V| = 27$. Now Burnside's lemma shows that there are 21 distinct colourings.

Exercise 3c

1. Using Burnside's lemma, check the solutions to questions 2 and 3 in exercise 3a.

2. Show that there are eight necklaces consisting of four beads, two of which are red and two blue.

3. A noughts-and-crosses board is filled with five Xs and four Os. How many different boards are there if rotations are counted as equivalent? What if both rotations and reflections are counted as equivalent?

4. Use Burnside's lemma to solve problem 3.5 which is about 2-colouring a 4 by 4 square grid.

5. (a) Find the number of ways to colour the 64 squares of a chessboard using k colours, with rotations counted as equivalent.

   (b) Repeat the calculation in (a) when reflections are also equivalent.

6. A rectangular flag, which is not square, consists of $2n$ vertical stripes of equal width, each of which can be one of k colours. How many such flags are there? What if it consists of $2n + 1$ stripes?

3.6 Permutations and symmetry

We now know that we can solve many colouring problems if we can calculate the sizes of invariant sets. With small groups this is not too difficult, but as the groups get more complicated we must develop a more powerful approach. It turns out that there is a strong relationship between symmetry groups and permutations, and we now explore this.

Before we can do this, we need to refine what we mean by permutations. Up to now, we have treated them as rearrangements of the elements of a finite set, but now we modify this definition slightly. The new insight is that a permutation can be thought of as a one-to-one function from a set to itself.

For example, we can define a permutation π on the set $\{1, 2, 3, 4, 5, 6\}$ as follows:

$$\begin{array}{lll} \pi(1) = 3 & \pi(2) = 1 & \pi(3) = 5 \\ \pi(4) = 4 & \pi(5) = 6 & \pi(6) = 2 \end{array}$$

To represent the permutation π, we 'follow through' what happens to the numbers. We see that 1 goes to 3, 3 goes to 5, 5 goes to 6, 6 goes to 2 and 2 goes to 1, which is where we started. So these five numbers form a *cycle* which we write as $(1\,3\,5\,6\,2)$. The number 4 is mapped to itself, so it is in a cycle on its own. Hence we write π in cycle notation as $(1\,3\,5\,6\,2)(4)$. Often we will omit the final cycle (4) and write this as $(1\,3\,5\,6\,2)$. It should now be obvious that every permutation can be written as a product of disjoint cycles.

The identity permutation has the property that $I(x) = x$ for all x, and the inverse of a permutation has the effect of reversing it. This gives us $\pi^{-1} = (2\,6\,5\,3\,1)$, or $(1\,2\,6\,5\,3)$ working through the cycle backwards. The inverse of $(1\,2\,3)(4\,5\,6)$ can be written $(3\,2\,1)(6\,5\,4)$ or $(1\,3\,2)(4\,6\,5)$.

As permutations are functions, they obey the associative law. It follows that the set of permutations of n elements, which is written S_n, forms a group of order $n!$. It is known as the symmetric group on n symbols.

Note also that *disjoint* cycles (which do not have any elements in common) commute; for example we have $(1\,5)(2\,3)(4\,6) = (4\,6)(2\,3)(1\,5)$. It is important, however, that the cycles *are* disjoint. Let us see what happens when we compose cycles which share elements.

Let us consider combining $S = (1\,2\,3\,4)$ and $T = (1\,2\,4)$ in two different ways, recalling that ST means 'do T and then do S'. To write ST in cycle notation, we first see where 1 goes. It is sent to 2 by T, and this is sent to 3 by S. Therefore the product ST begins $(1\,3\,\cdots)$. Next we ask what happens to 3. It is ignored by T and then sent to 4 by S, so now we have $ST = (1\,3\,4\,\cdots)$. Now we pursue 4 — it is sent to 1 by T and then to 2 by S. Finally 2 goes to 4 under T and then to 1 under S. As a result we have $(1\,2\,3\,4)(1\,2\,4) = (1\,3\,4\,2)$. On the other hand, if we consider $(1\,2\,4)(1\,2\,3\,4)$, we reason as follows: 1 goes to 2 and then to 4; 4 goes to 1 and then 2; 2 goes to 3 and no further; 3 goes to 4 and then to 1. Hence $(1\,2\,4)(1\,2\,3\,4) = (1\,4\,2\,3)$.

These are different permutations, so products of cycles which share elements do not commute. In view of this, we will almost always split permutations into disjoint cycles.

The next problem shows how permutations can be used to help understand symmetries.

Problem 3.6

(a) How many rotational symmetries does a regular tetrahedron have?

(b) Express these symmetries as permutations of the vertices.

(c) What are the possible combined effects of two 120° rotations around different axes?

We label the vertices of the tetrahedron as shown in figure 3.9.

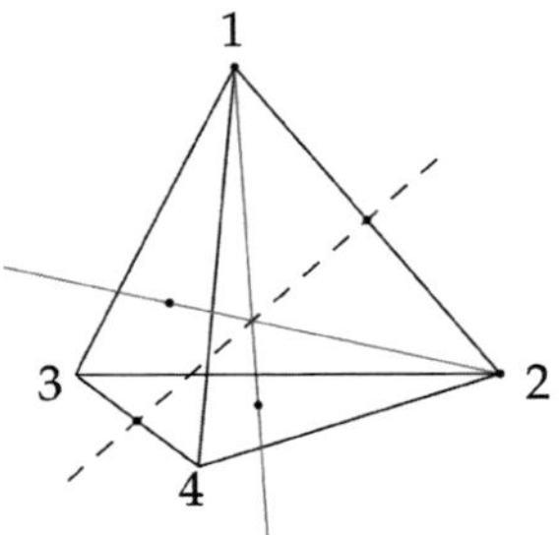

Figure 3.9

For part (a) we can imagine first moving vertex 1 into any of four different positions, and choosing one of three possible orientations for the opposite face to give a total of twelve symmetries.

For (b) we note that there are two types of rotational symmetry available. One type rotates by 120° or 240° about an axis through a vertex and the centre of the opposite face (shown in grey in figure 3.9). These eight symmetries correspond to the following permutations of the vertices:

$$(2\,3\,4), (2\,4\,3), (1\,3\,4), (1\,4\,3), (1\,2\,4), (1\,4\,2), (1\,2\,3), (1\,3\,2).$$

The other symmetries are half-turns around axes which pass through the centres of opposite edges (the dashed line in figure 3.9). These correspond to the permutations $(1\,2)(3\,4), (1\,3)(2\,4), (1\,4)(2\,3)$.

Finally we must not forget the identity symmetry to get a total of twelve as expected.

In part (c) we are asked to consider, for example, the effect of rotation 120° around the axis through 1, and then 120° around the axis through 2. Visualising this is not easy, but the result can readily be found using permutations.

Without loss of generality, the first rotation corresponds to $(2\,3\,4)$. The second corresponds to $(1\,4\,3)$ or $(1\,3\,4)$ depending on which way we rotate. We now simply compute the two products $(1\,4\,3)(2\,3\,4) = (1\,4\,2)$ and $(1\,3\,4)(2\,3\,4) = (1\,3)(2\,4)$.

These show that the result of the two rotations can either be another 120° rotation around a new axis, or a 180° rotation.

Let us now return to a hexagon with its vertices labelled from 1 to 6, as in diagram A in figure 3.10.

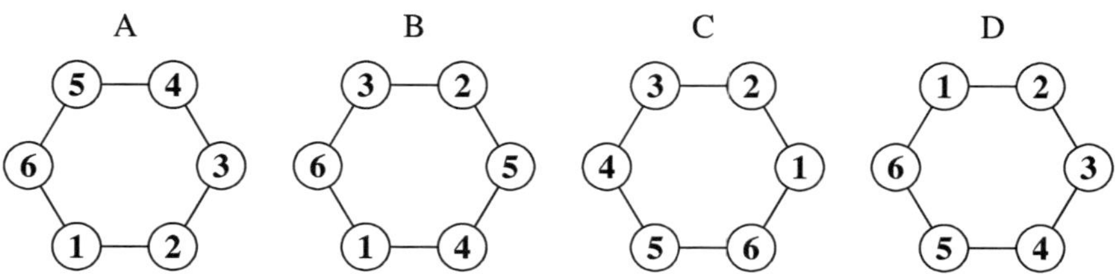

Figure 3.10

The hexagons B, C and D can be thought of as relabelled versions of A. Hexagon C is obtained from A by moving the label 1 to where label 3 used to be, 2 to where 4 was, and so on, and we will consider that as the permutation $(1\,3\,5)(2\,4\,6)$. Similarly the permutation $(2\,4)(3\,5)$ takes A to B and $(1\,5)(2\,4)$ takes A to D.

However, $(2\,4)(3\,5)$ does not correspond to any physical transformation. It is impossible to rotate or reflect A in any way to achieve B. On the other hand, $(1\,3\,5)(2\,4\,6)$ is a rotation of 120° about the centre of the hexagon and $(1\,5)(2\,4)$ is a reflection in the line through 3 and 6.

This is saying that the two groups introduced earlier — C_6 and D_6 — can be thought of as subgroups of S_6. The cyclic group C_6 was defined as $\{I, R, R^2, R^3, R^4, R^5\}$, but these can be expressed as the set of permutations $\{I, (1\,2\,3\,4\,5\,6), (1\,3\,5)(2\,4\,6), (1\,4)(2\,5)(3\,6), (1\,5\,3)(2\,6\,4), (1\,6\,5\,4\,3\,2)\}$.

One could argue that these groups are not identical since their elements are defined differently, but they have exactly the same structure, so we will refer to them as being the same group. Note that we could also think of the group of rotations — that is, C_6 — as permuting the sides or even the diagonals of the hexagon. This may seem like an unimportant observation, but it is going to become more substantial when we define the group of transformations of a polyhedron and then reinterpret it in terms of permutations of the vertices, edges, faces or diagonals.

In the same way, the dihedral group D_6 corresponds to the group of permutations of the vertices

$$I, (1\,2\,3\,4\,5\,6), (1\,3\,5)(2\,4\,6), (1\,4)(2\,5)(3\,6), (1\,5\,3)(2\,6\,4), (1\,6\,5\,4\,3\,2),$$
$$(1\,2)(3\,6)(4\,5), (1\,3)(4\,6), (1\,2)(3\,6)(4\,5), (1\,5)(2\,4), (1\,6)(2\,5)(3\,4), (3\,5)(2\,6)$$

where the permutations on the second row represent the reflections.

The next problem shows how thinking in terms of permutations can help us count colourings.

Problem 3.7

How many 3-colourings do the vertices of a regular hexagon have under D_6?

The problem of counting colourings under symmetries such as rotations and reflection has been reduced, by Burnside's lemma, to that of counting invariant sets. In section 3.5 and exercise 3c, this is done 'by hand', since it is easy enough to work out how many of the raw colourings of a configuration are preserved by a particular rotation or reflection. When the set of colourings becomes much larger and the symmetries more complex, this becomes harder. However, it turns out that there is a simple (and mechanical) way to do this once the transformations in question have been expressed as permutations in cycle notation.

To solve this problem we will, for example need to know how many colourings are invariant under a rotation of 120°. It is obvious that all we require is that alternate vertices have the same colour, so there are nine colourings in the invariant set. A rotation of 120° can be expressed as the permutation of the form $(1\,3\,5)(2\,4\,6)$, which consists of two cycles of length 3. If a colouring is fixed by this permutation, vertices inside the same cycle will have the same colour. The two cycles are independent and we must choose one of three possible colours for each cycle, so it turns out that $|\Omega_{(1\,3\,5)(2\,4\,6)}| = 3 \times 3$. This is, admittedly, exactly the 'hands on' argument, but if we focus on the cycle structure of the permutation we see that the two important things are the number of cycles and the number of colours.

To solve the problem we can summarise what we know about the group D_6 as shown in table 3.5 on the following page. We assume that the letters in the permutations represent different numbers. The rows Reflection (M) and Reflection (V) refer to reflections whose axes pass through midpoints of opposite sides and opposite vertices respectively.

Now Burnside's lemma shows that the number of colourings is

$$\frac{1}{12}(1 \times 729 + 2 \times 3 + 2 \times 9 + 1 \times 27 + 3 \times 81 + 3 \times 27) = 92.$$

We must be careful to include the singletons (e) and (f) in the last line of the table to remind ourselves that to get an invariant colouring under that symmetry we make four independent colour choices, even though it is natural to express the symmetry as a product of two cycles.

Type of symmetry	Number	Permutation type	$\lvert\Omega_\pi\rvert$
Identity	1	$(a)(b)(c)(d)(e)(f)$	3^6
60° rotation	2	$(a\,b\,c\,d\,e\,f)$	3
120° rotation	2	$(a\,b\,c)(d\,e\,f)$	3^2
180° rotation	1	$(a\,b)(c\,d)(e\,f)$	3^3
Reflection (M)	3	$(a\,b)(c\,d)(e\,f)$	3^3
Reflection (V)	3	$(a\,b)(c\,d)(e)(f)$	3^4

Table 3.5

Exercise 3d

1. By labelling the vertices of an equilateral triangle, interpret D_3 as a group of permutations in S_3.

2. By labelling the vertices of a square, interpret D_4 as a group of permutations in S_4.

3. By labelling the edges of a regular tetrahedron, interpret its rotational symmetries as a group of permutations in S_6.

4. How many permutations π of the set $\{1, 2, \ldots, 2n\}$ with no fixed points have the property that $\pi^2 = I$?

5. An altitude is drawn on each face of a regular tetrahedron. How many distinguishable arrangements are possible?

3.7 The cycle index

Our approach to the previous problem motivates the following definition:

CYCLE TYPE

A permutation consisting of n_1 cycles of length 1, n_2 cycles of length 2, and so on up to n_r cycles of length r has cycle type $x_1^{n_1} x_2^{n_2} \ldots x_r^{n_r}$.

As an example we can return yet again to D_6. There we had the following symmetries:

(i) the identity permutation: cycle type x_1^6;

(ii) two permutations of the form $(a\,b\,c\,d\,e\,f)$: cycle type x_6^1;

(iii) two permutations of the form $(a\,b\,c)(d\,e\,f)$: cycle type x_3^2;

(iv) four permutations of the form $(a\,b)(c\,d)(e\,f)$: cycle type x_2^3;

(v) three permutations of the form $(a\,b)(c\,d)(e)(f)$: cycle type $x_1^2x_2^2$.

Note that if we multiply together the subscripts and indices for each variable and then add, we always obtain 6, which is the length of the permutation. This is the reason for including the 'superfluous' indices which are 1.

We can use the cycle index notation to streamline our solution to problem 3.7.

A cycle type, such as $x_1^{n_1}x_2^{n_2}\ldots x_r^{n_r}$, looks rather like a term in a polynomial with several variables $x_1, x_2^{n_2}, \ldots, x_r^{n_r}$. In the case of the cube diagonals, the polynomial has four variables. The sizes of the invariant sets were calculated by substituting 3 (the number of colours) for the variables in each cycle type. For example, the permutations of the form $(a\,b\,c)(d\,e\,f)$, which have cycle type x_3^2, correspond to 120° rotations. We dealt with these by noting that the colours of the vertices must alternate so we have free choices as to the colour of two adjacent vertices and the rest is determined. This gave nine colourings in the invariant set, and this calculation is achieved by substituting 3 for x_2 in the cycle type.

We next observed that there were two permutations of the cycle type x_3^2, so it was necessary to multiply by 2 in calculating the size of the invariant set corresponding to all permutations of this type . This explains the coefficient in the polynomial term $2x_3^2$. Finally, all of these terms are added to create a polynomial in four variables, namely

$$P_{D_6}(x_1, x_2, x_3, x_6) = \frac{1}{12}(x_1^6 + 3x_1^2x_2^2 + 4x_2^3 + 2x_3^2 + 2x_6).$$

This serves as a sort of code for the types of the permutations in the group. It is called the *cycle index* of the group of transformations. Eventually we equate all of the variables to the number of colours which are being used. In particular $P_{D_6}(3,3,3,3) = 92$.

Note that if we equate all the variables in a cycle index to 1, we obtain the value 1. That is a good check that we have not forgotten anything in constructing the polynomial; after all, there is only one way to colour something using only one colour.

The whole point of this, of course, is that Burnside's lemma can be restated as follows:

BURNSIDE'S LEMMA: CYCLE INDEX FORM

The number of k-colourings under the permutation group G which has cycle index $P_G(x_1, \ldots, x_r)$ is given by $P_G(k, \ldots, k)$.

Problem 3.8

An equilateral triangle is divided into six regions by the medians. Find the number of ways to colour the regions using k colours.

The group is D_3. Number the regions 1 to 6 anticlockwise. The permutations are:

(i) identity permutation: cycle type x_1^6;

(ii) two rotations like $(1\,3\,5)(2\,4\,6)$: cycle type x_3^4;

(iii) three reflections like $(1\,2)(3\,6)(4\,5)$: cycle type x_2^3.

Hence the cycle index is $\frac{1}{6}(x_1^6 + 3x_2^3 + 2x_3^2)$. The number of colourings is $\frac{1}{6}k^2(k^4 + 3k + 2)$. For three colours there would be 138 colourings.

3.8 Symmetries of a cube

Now we turn to a more demanding three-dimensional problem.

Problem 3.9

In how many ways can

(a) the vertices;

(b) the faces;

(c) the edges

of a cube be coloured using k colours?

The first thing we need to do is to understand the rotational symmetries of a cube. Then we will see how these rotations affect vertices, faces and edges in order to construct the relevant cycle indexes.

Our first observation is that if we wish to rotate a cube in space, we can start by choosing a specific face and putting it into one of six positions. Once we have chosen that position we can choose the orientation of the face in one of four ways so we expect a total of twenty-four symmetries.

Apart from the identity there are four types of symmetry to consider:

(i) rotation by 120° around an axis joining opposite corners. There are four axes to choose from and the rotations can be clockwise or anticlockwise giving eight in all.

(ii) rotation by 90° around an axis through the centres of opposite faces. There are three axes to choose from and two directions, giving six in all.

(iii) rotations by 180° around an axis through the centres of opposite faces. There is one such rotation for each of the three axes.

(iv) rotation by 180° around an axis through the centres of opposite edges. A cube has twelve edges so there are six axes to choose from and one possible rotation per axis.

It is reassuring that $1+8+6+3+6=24$ as expected.

Now we interpret each type of rotation as a permutation of (a) vertices, (b) faces and (c) edges of the cube and find its cycle type.

(i) The rotation takes figure 3.11 to figure 3.12.

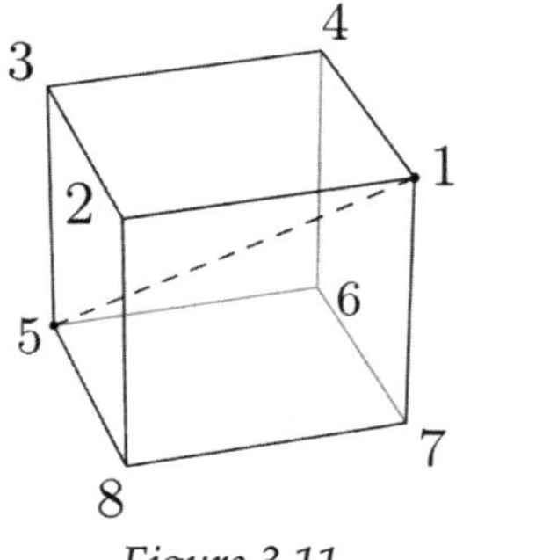

Figure 3.11

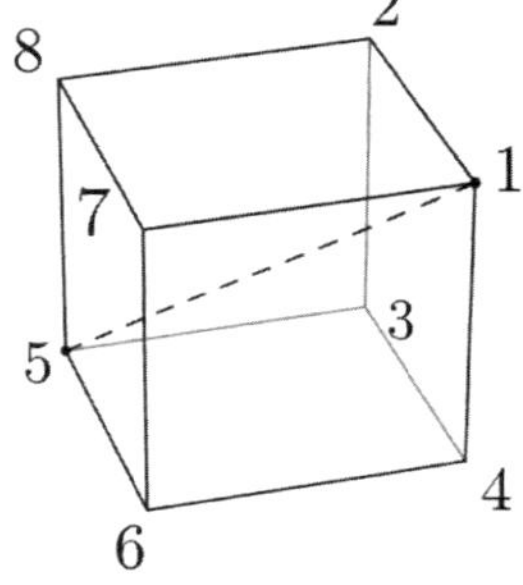

Figure 3.12

Now we find the required cycle types.

(a) Vertices 1 and 5 are fixed and we have the three cycles $(2\,4\,7)$ and $(3\,6\,8)$. Thus the cycle type is $x_1^2 x_3^2$.

(b) The three faces including vertex 1 form a 3-cycle, as do the three touching vertex 5. Thus the cycle type is x_3^2.

(c) The three edges including vertex 1 and the three touching vertex 5 form cycles. The remaining six edges form two more 3-cycles. Thus the cycle type is x_3^4.

(ii) The rotation takes figure 3.13 to figure 3.14.

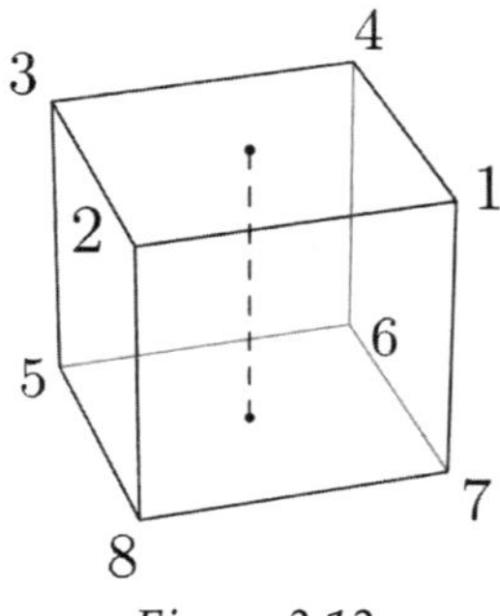

Figure 3.13

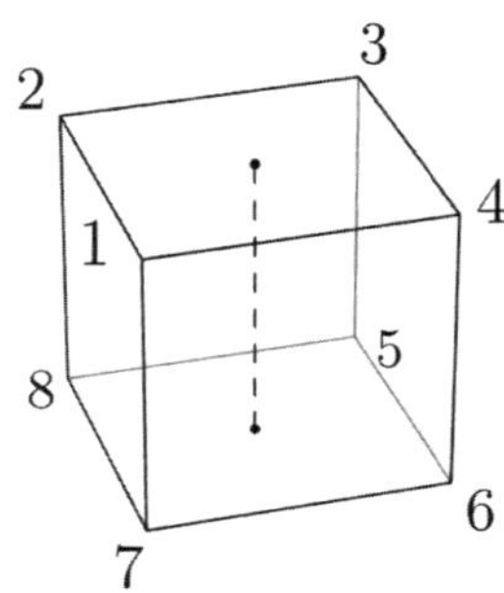

Figure 3.14

Now we find the required cycle types.

(a) The vertices form the cycles $(1\,2\,3\,4)$ and $(5\,6\,7\,8)$. Thus the cycle type is x_4^2.

(b) The two faces perpendicular to the axis are fixed and the remaining four form a cycle. Thus the cycle type is $x_1^2 x_4$.

(c) The four edges bounding each fixed face form a cycle as do the four parallel to the axis. Thus the cycle type is x_4^3.

(iii) The rotation takes figure 3.15 to figure 3.16.

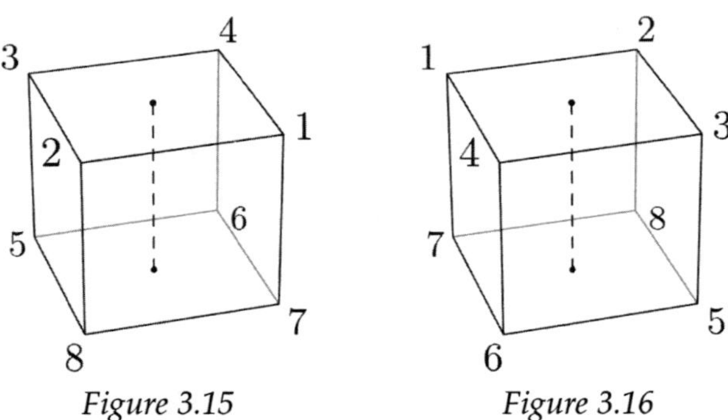

Figure 3.15 *Figure 3.16*

Now we find the required cycle types.

(a) Each of the 4-cycles in rotation type (ii) splits into two 2-cycles. Thus the cycle type is x_2^4.

(b) The 4-cycle in the type (ii) rotation splits again. Thus the cycle type is $x_1^2 x_2^2$.

(c) Again, the 4-cycles split. Thus the cycle type is x_2^6.

(iv) The rotation takes figure 3.17 to figure 3.18.

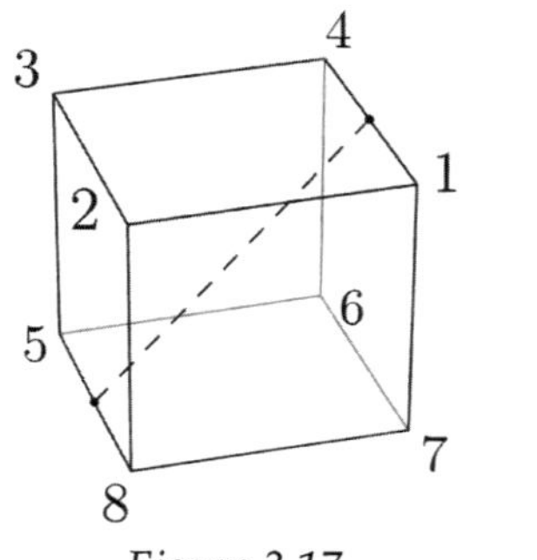

Figure 3.17

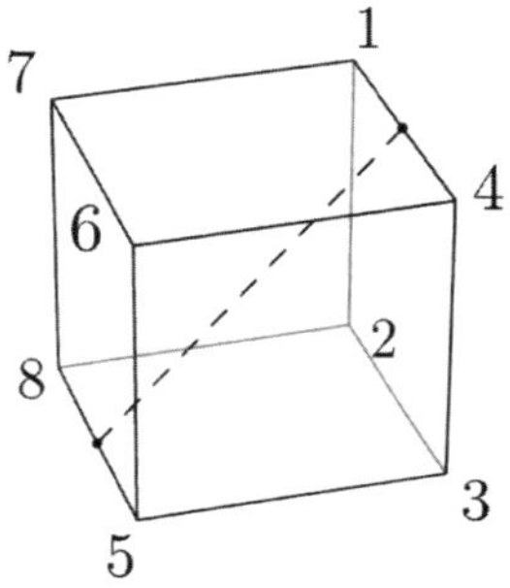

Figure 3.18

Now we find the required cycle types.

(a) Vertices at each end of the two fixed edges swap and the remaining four form two 2-cycles. Thus the cycle type is x_2^4.

(b) The two faces parallel to the axis swap as do each pair of faces which meet on a fixed edge. Thus the cycle type is x_2^3.

(c) The edges which intersect the axis are fixed. There are two more edges parallel to these and they swap places.

The two faces parallel to the axis swap places, so the four edges on the boundary of one of these faces swap with the four on the boundary of the other. This gives four more 2-cycles. Thus the cycle type is $x_1^2 x_2^5$.

We can summarise all this analysis in table 3.6.

It is now a straightforward matter to construct the required cycle indexes and substitute k for each variable. We discover that with k colours there are

(a) $\dfrac{6k^2 + 17k^4 + k^8}{24}$ vertex colourings;

(b) $\dfrac{8k^2 + 12k^3 + 3k^4 + k^6}{24}$ face colourings;

(c) $\dfrac{6k^3 + 8k^4 + 3k^6 + 6k^7 + k^{12}}{24}$ edge colourings.

Rotation type	Number	Vertex cycle type	Face cycle type	Edge cycle type
(i) - (120°)	8	$x_1^2x_3^2$	x_3^2	x_3^4
(ii) - (90°)	6	x_4^2	$x_1^2x_4$	x_4^3
(iii) - (180°)	3	x_2^4	$x_1^2x_2^2$	x_2^6
(iv) - (180°)	6	x_2^4	x_2^3	$x_1^2x_2^5$
I - (0°)	1	x_1^8	x_1^6	x_1^{12}

Table 3.6

Problem 3.10

Prove that if we interpret the rotational symmetries of a cube as permutations of its four diagonals we obtain the group S_4.

One approach to this problem would be to go through the four types of rotational symmetry of a cube and see how each of these affects the diagonals. However, this turns out to be unnecessary. We already know that the cube has 24 rotational symmetries and that S_4 has order $4! = 24$. It is also clear that rotations of the cube permute the diagonals, so the only way we could fail to get the whole of S_4 would be if there were two different rotations of the cube, say σ and π, which gave rise to the same permutation of the diagonals.

We assume, for contradiction that these rotations exist and consider the rotation $\sigma\pi^{-1}$. This is not the identity since $\sigma \neq \pi$ but it does fix all the diagonals of the cube. In particular vertex 1 is sent either to 1 or to 5, and we consider these cases in turn.

First we assume 1 is sent to 1.

Since the diagonal (26) is fixed, 2 is sent to 2 or 6, but, since 6 is not adjacent to 1, we see that 2 must go to 2 or else the edge (12) would be destroyed by the rotation. The same reasoning applies to the other vertices adjacent to 1, so 4 and 7 are also fixed,which is enough to fix the whole cube. This contradicts the initial assumption that $\sigma\pi^{-1}$ is not the identity.

We are left with the case where 1 is sent to 5. Reasoning as before we see that 2 is sent to 6, 4 to 8 and 7 to 3. The cube before and after this transformation is shown in figure 3.19.

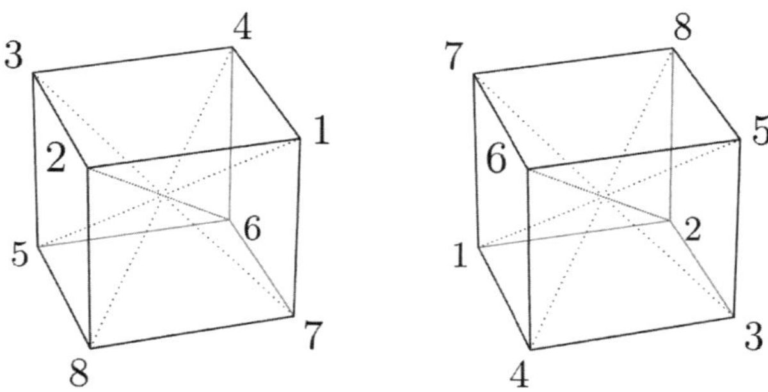

Figure 3.19

This transformation is a symmetry of the cube called *reflection in the centre*. However it is not a rotation. One way to see this is to note that its cycle type is x_2^4 which does not occur in the cycle index for rotational symmetries of the cube. An alternative is to imagine holding the cube with vertex 1 in the palm of your hand and pointing your thumb, index finger and middle finger in the directions of vertices 2, 4 and 7 respectively. Before the reflection in the centre this requires a right hand, and afterwards it requires a left hand, so the *orientation* of the three edges has been reversed, which is impossible for a rotation (and always happens with a reflection).

Exercise 3e

1. Find the vertex, edge and face cycle indexes for a regular octahedron.

2. An arrow is drawn on each edge of a cube. In how many ways can this be done?

3. A diagonal is drawn on each faces of a cube. In how many ways can this be done?

4. The standard patterns of 1 to 6 dots are added to faces of a cube. In how many ways can this be done if
   (a) there are no restrictions;
   (b) the dots on opposite faces must sum to seven?

3.9 Counting collineations

Problem 3.11

In how many ways can the Fano plane be coloured using four colours?

You met the Fano plane in section 1.6 (on page 33) where it featured in the discussion of Kirkman's schoolgirl problem. We need to discuss the symmetries of this configuration and will do so by analysing the permutations of its seven points. Clearly there are 7! permutations altogether, but these are not all acceptable as symmetries.

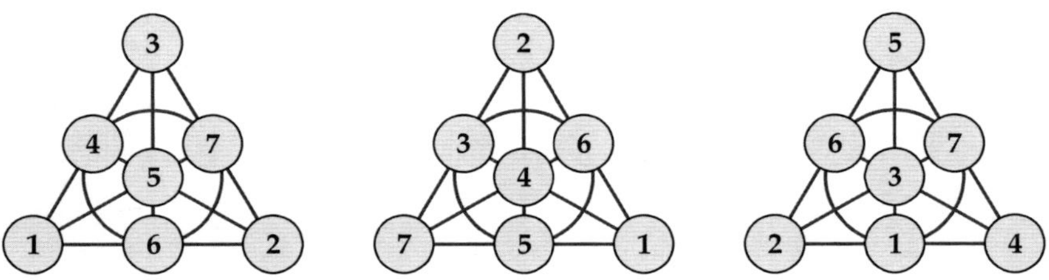

Figure 3.20

The diagram on the left in figure 3.20 shows the Fano plane labelled as it is in chapter 1. The other two diagrams are relabellings of the points, which can be considered as permutations of seven elements. The middle diagram has the same seven lines as the first, namely 126, 134, 157, 237, 245, 356 and 467. The third diagram has different lines including, for example 124. So, whereas the first two diagrams can be thought of as 'identical' in the sense that the lines and points have been renamed consistently, the third is to be considered as a different plane altogether. The permutation which takes us from the first to the third diagram does not describe a symmetry of the plane. We will be restricting ourselves to permutations, such as $(1\,2\,3\,4\,5\,6\,7)$ shown in the middle diagram, which retain the line structure. These are known as *collineations*.

We begin by determining how many collineations there are. To construct one we send 1 to one of seven places and 2 to one of six. These choices determine the position of 6 since 126 is a line. At this point 3 can go to one of four places. Now the collinearities 134, 356 and 237 determine the positions of 4, 5 and 7 respectively.

This means there are at most $7 \times 6 \times 4 = 168$ potential collineations. However, we must check that these all preserve the last three lines, namely 157, 245 and 467. Fortunately this always happens. For example, the points 5 and 7 define a line $a57$ and since 7 is not on the lines 245 or 356, we see that a cannot be 2, 3, 4 or 6. The argument for the other lines is similar, so there are indeed 168 collineations. More generally we have the following key fact.

A collineation is uniquely determined by the positions of three non-collinear points.

Our next task is to find the cycle-types of the collineations. We will do this systematically by considering the number of fixed points in a collineation.

Case 1: At least four fixed points.

Four fixed points include three non-collinear ones so our key fact now shows that the collineation must be the identity. Thus we have one permutation of cycle type x_1^7.

Case 2: Exactly three fixed points.

The three points must be collinear. We assume, without loss of generality, that they are 126. The point 3 must now move and can go to one of three places. The options are shown in figure 3.21. Completing the collineation using the lines 134, 356 and 237 as usual, we see that the three permutations are $(3\,4)(5\,7)$, $(3\,5)(4\,7)$ and $(3\,7)(4\,5)$. Since the fixed line could have been chosen in seven ways we see that there are 21 collineations with cycle type $x_1^3x_2^2$.

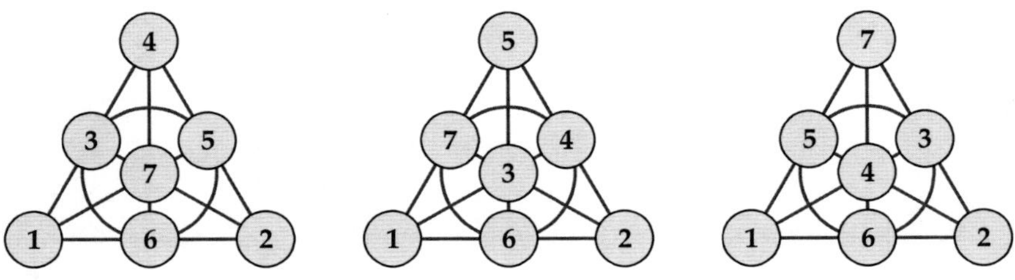

Figure 3.21

A collineation cannot have exactly two fixed points since the third point on the line they define would also need to be fixed.

Case 3: Exactly one fixed point.

We assume that 1 is fixed and ask whether the permutation contains a 2-cycle (also called a *transposition*).

Case 3A: Exactly one fixed point and a transposition.

The points which swap must define a line through 1. We assume 2 and 6 swap. Now the position of 3 is enough to determine the collineation. If 3 goes to 4 then 5 and 7 are fixed and the collineation has been counted in case 2. The situations when 3 goes to 5 and 7 are shown in figure 3.22.

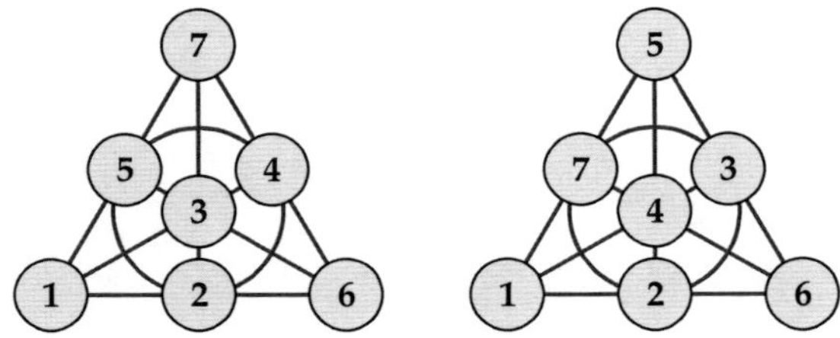

Figure 3.22

They correspond to the permutations $(2\,6)(3\,5\,4\,7)$ and $(2\,6)(3\,7\,4\,5)$.

We could have chosen the fixed point in one of seven ways and the transposition in one of three, so we have $7 \times 3 \times 2 = 42$ collineations of cycle type $x_1x_2x_4$.

Case 3B: Exactly one fixed point and no transposition.

In this case the cycle type must be $x_1x_3^2$. We ask whether the points in the 3-cycles are collinear. It cannot be that both 3-cycles permute lines, since any two lines intersect. We must therefore have at least one 3-cycle of non-collinear points.

Without making any assumptions about the fixed point, we may assume we have the 3-cycle (1 2 3). Our key fact shows that this determines the collineation. It is shown in figure 3.23 and corresponds to $(1\,2\,3)(4\,6\,7)$. To count these permutations we must count the non-collinear triples *abc*. There are seven choices for *a*, six for *b* and four for *c*, but this counts each triple three times since the smallest number appears as each of *a*, *b* and *c*.

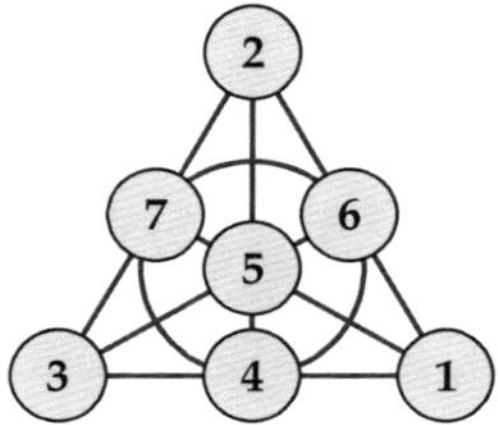

Figure 3.23

Thus we have $7 \times 4 \div 3 = 56$ collineations of cycle type $x_1 x_3^2$

Case 4: No fixed points.

Since transpositions give rise to fixed points, the only possible cycle types are $x_3 x_4$ and x_7. We consider these in turn.

If a collineation contains a 3-cycle of non-collinear points then our work in case 3b shows it must have a fixed point. We may therefore assume that the three cycle is $(1\,2\,6)$. There are now three choices of where to send 3. These are shown in figure 3.24 but it easy to see that each has a fixed point (shown in heavy grey). Thus there are no collineations of cycle type $x_3 x_4$.

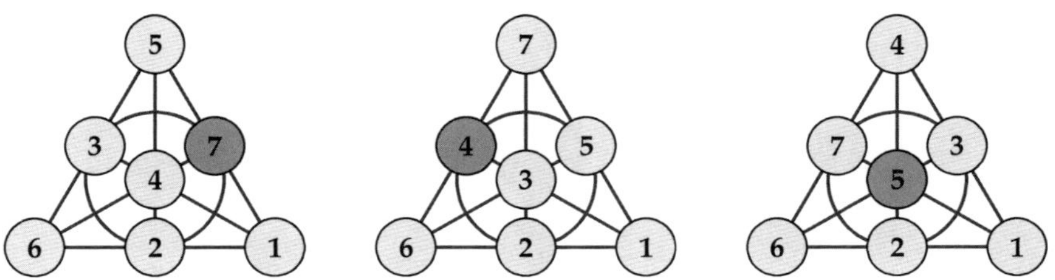

Figure 3.24

Counting the collineations of cycle type x_7 is delicate, but fortunately not necessary. We know there must be $168 - 1 - 21 - 42 - 56 = 48$ of them.

Summarising, we have

- the identity, which is of cycle type x_1^7;
- 21 permutations of cycle type $x_1^3 x_2^2$;
- 42 permutations of cycle type $x_1 x_2 x_4$;
- 56 permutations of cycle type $x_1^1 x_3^2$;
- 48 permutations of cycle type x_7^1.

Hence the cycle index is $\frac{1}{128}(x_1^7 + 21x_1^3x_2^2 + 42x_1^1x_2^1x_4^1 + 56x_1^1x_3^2 + 48x_7^1)$. The number of k-colourings is $\frac{1}{168}k(k^6 + 21k^4 + 98k^2 + 48)$ and for $k = 4$ this is 42.

3.10 Counting graph colourings

Problem 3.12

How many graphs are there with four vertices?

You met this problem in question 2 of exercise 2a where the eleven such graphs are listed. Recall that the vertices of a graph are unlabelled and so treated as indistinguishable. If the vertices were labelled, this problem would be easier. There are 6 edges and each can be 'coloured' as either 'present' or 'absent', and so there are 64 different graphs. In general, there are $2^{\frac{n(n-1)}{2}}$ labelled graphs of order n. The difference between this and the problem where the vertices are unlabelled is that graphs which look superficially different turn out to be the same.

However, the method of colouring edges as 'present' or 'absent' is the key to solving this problem. We begin with K_4, the complete graph with four vertices, and we have to count the two-colourings of the edges of this graph. Clearly Burnside's lemma and the cycle index will turn out to be useful in doing this.

We begin by identifying the group of permutations. Since there are 4 vertices, which can be permuted in any way, this is the symmetric group

S_4 which has order 24. We need to find the cycle index when this group is applied to the set of edges of K_4. It turns out that it is useful to begin by considering the cycle index of the vertices under S_4, which has already been calculated as

$$P^V_{S_4}(x_1, x_2, x_3, x_4) = \frac{1}{24}(x_1^4 + 6x_1^2x_2^1 + 8x_1^1x_3^1 + 3x_2^2 + 6x_4^1)$$

but now includes a superscript V to indicate that we are permuting the set of vertices. We now have to calculate the cycle index of the edges under S_4. We can write each edge in terms of the two vertices it joins and use this to 'translate' the permutations of vertices into permutations of edges. We do this by looking at the different cycle types.

Vertex permutation	Cycle type	Edge permutation	Cycle type
(1)(2)(3)(4)	x_1^4	(12)(13)(14)(23)(24)(34)	x_1^6
(1 2 3 4)	x_4^1	(12 23 34 14)(13 24)	$x_2^1x_4^1$
(1 3)(2 4)	x_2^2	(13)(12)(12 34)(14 23)	$x_1^2x_2^2$
(1 2 3)(4)	$x_1^1x_3^1$	(12 23 13)(14 24 34)	x_3^2
(1 2)(3)(4)	$x_1^2x_2^1$	(12)(13 23)(14 24)(34)	$x_1^2x_2^2$

Table 3.7

To see what happens to a particular edge, such as 12, we look at the vertices at its ends. Under the permutation (1 3)(2 4), 1 becomes 3 and 2 becomes 4, so 12 becomes 34. Under the permutation (1 2 3)(4), 12 becomes 23. Remember, of course, that 21 is the same edge as 12.

The translating results in table 3.7 and we find that the edge cycle index is

$$P^E_{S_4}(x_1, x_2, x_3, x_4) = \frac{1}{24}(x_1^6 + 8x_3^2 + 6x_1^2x_2^2 + 3x_1^2x_2^2 + 6x_2^1x_4^1).$$

Now we evaluate this polynomial with $x_1 = x_2 = x_3 = x_4 = 2$ to obtain the correct answer of 11.

Exercise 3f

1. Find the number of ways to colour the vertices of a regular tetrahedron using k colours.

2. Find the number of ways to colour the edges of a regular tetrahedron using k colours.

3. Find the number of ways to colour the eight beads of a necklace using k colours.

4. Find the number of graphs of order 5.

3.11 Pólya enumeration

The technique of using Burnside's lemma together with the cycle index is known as *Pólya enumeration,* after the Hungarian mathematician George Pólya who worked on combinatorics and number theory as well as studying the heuristics of problem-solving. It incorporates a generalisation which allows us to solve problems about patterns in which the colours are restricted in some way.

Problem 3.13

An equilateral triangle is divided into six regions by the medians. Find the number of ways to colour the regions so that two are coloured blue, two are coloured red and two are coloured green.

In the solution to problem 3.8 on page 99 we showed that the relevant cycle index is given by $P(x_1, x_2, x_3) = \frac{1}{6}(x_1^6 + 3x_2^3 + 2x_3^2)$. Let us re-examine the logic behind the claim that $P(k, k, k)$ is the number of *unrestricted* colourings of the triangle regions.

We are counting the number of invariant sets for a given permutation. Permutations composed of three cycles, like $(1\,2)(3\,6)(4\,5)$, give rise to k^3 invariant sets because we must choose a colour for each cycle. Hence the three permutations of type x_2^3 produce $3k^3$ invariant sets. Summing over the cycle types and dividing by 6, we obtain, by Burnside's lemma, the number of colourings. This is equivalent to setting $x_1 = x_2 = x_3 = k$ in the cycle index.

Suppose, however, that we substitute $x_1 = b + g + r, x_2 = b^2 + g^2 + r^2$ and $x_3 = b^3 + g^3 + r^3$. The result is a polynomial of degree 6 in three variables, and it turns out that the answer to problem 3.13 is given by the coefficient of $b^2g^2r^2$ when this polynomial is expanded.

First we consider x_2^3 which becomes $(b^2 + g^2 + r^2)^3$. This expands to give

$$b^6 + g^6 + r^6 + 3b^4g^2 + 3b^4r^2 + 3g^4b^2 + 3g^4r^2 + 3r^4b^2 + 3r^4g^2 + 6b^2g^2r^2.$$

There are 27 terms in this expression if we split them up so that all the coefficients are 1. The first corresponds to choosing b^2 in each bracket, which in turn corresponds to the colouring where the regions in each cycle are blue. In the same way, b^4g^2 terms correspond to colourings where two 2-cycles are blue and the third green, while $b^2g^2r^2$ terms correspond to colourings where the three 2-cycles are all different colours. In particular, the fact that the coefficient of $b^2g^2r^2$ in $(b^2 + g^2 + r^2)^3$ is 6 tells us that a transformation with cycle type x_2^3 (that is, a reflection) has exactly 6 invariant colourings with two blue, two red and two green regions.

Now we consider x_1^6 which becomes $(b + g + r)^6$. The coefficient of $b^2g^2r^2$ when this is expanded is given by $\binom{6}{2\,2\,2} = 90$. This corresponds to the fact that we must choose b from two brackets, r from two brackets and g from two brackets. This in turn corresponds to the fact that to form a colouring with two regions of each colour we simply choose two blue, two green and two red regions. The permutation with cycle type x_1^6 is the identity, so all 90 colourings we have counted are fixed by it.

Finally we consider x_3^2 which becomes $(b^3 + g^3 + r^3)^2$. Here there are no $b^2g^2r^2$ terms which corresponds to the fact that there are two ways to assign colours to two 3-cycles using blue, red and green twice each. In particular, invariant colourings for rotations of the triangle cannot have this colour scheme.

Burnside's lemma tells us to take the average size of the relevant invariant sets over all permutations, so the solution to the problem is given by $\frac{1}{6}(90 + 3 \times 6 + 2 \times 0) = 18$.

This is precisely the coefficient of $b^2g^2r^2$ obtained when the cycle index $P = \frac{1}{6}(x_1^6 + 3x_2^3 + 2x_3^2)$ is evaluated at $x_i = b^i + g^i + r^i$ for $i = 1, 2, 3$.

By a similar process, we can construct table 3.8, in which a colour pattern such as $\{3,2,1\}$ is taken as meaning all configurations with three regions of one colour, two of another and one of a third.

Colour pattern	Number of colourings
$\{6,0,0\}$	3
$\{5,1,0\}$	6
$\{4,2,0\}$	24
$\{4,1,1\}$	15
$\{3,3,0\}$	12
$\{3,2,1\}$	60
$\{2,2,2\}$	18

Table 3.8

The same technique can be applied to problem 3.12 to classify the graphs of order four according to the number of edges. Here we are 'colouring' the edges as present or absent and, rather than do this with a and b, it is simpler to use x and 1. The cycle index for edges becomes

$$P^E_{S_4} = \frac{1}{24}\Big((x+1)^6 + 8(x^3+1)^2 + 6(x+1)^2(x^2+1)^2 + 3(x+1)^2(x^2+1)^2 + 6(x^2+1)^1(x^4+1)\Big)$$

and this is expanded as $P(x) = x^6 + x^5 + 2x^4 + 3x^3 + 2x^2 + x + 1$. This is interpreted as the fact that there is one graph of order four with six edges, one with five, two with four, three with three, two with two, one with one and one with no edges.

This would not have been particularly difficult to achieve by hand, but when the number of vertices increases, the number of such graphs increases fast and a systematic method is needed. It is true that the expansion of the cycle index becomes much more taxing, but there are free programs on the web which will expand brackets. The really difficult problem is finding a foolproof method of counting, and this is the strength of Pólya enumeration.

Exercise 3g

1. Find the number of ways to colour the vertices of a regular pentagon so that two are blue, two red and one green.

2. Find the number of ways to colour the faces of a cube so that there are equal numbers of blue, red and green faces.

3. Find the number of ways to colour the edges of a cube so that there are equal numbers of blue, red and green edges.

4. Classify the graphs of order five according to the number of edges.

Chapter 4

Partitions

A partition is a division of something into parts. If this is carried out in accordance with some rule, then it is often worth investigating the number of ways in which it can be done.

In this chapter, we examine three different kinds of partitioning, as applied to positive integers, to sets and to permutations.

4.1 Partitioning integers

A *partition* of a positive integer is a way of expressing it as a sum of positive integers, where the order of parts does not matter.

Partitions should not be confused with *compositions*, where the order is important. It is true that $12 = 5 + 2 + 3 + 2 = 2 + 5 + 2 + 3$. These are different compositions of 12 but they count as the same partition. For that reason, it is customary to order the parts of a partition in descending order, when both these compositions would be written as 5 3 2 2.

A composition of twelve into four parts can be coded using twelve Xs separated by three †s which indicate where the breaks come. For example, 5 3 2 2 is coded XXXXX†XX†XXX†XX. There are 11 gaps between the Xs and the †s are inserted into three different gaps. Hence the number of distinct compositions of 12 into four parts is given by $\binom{11}{3}$ or 165, and the total number of compositions of 12 is 2^{11}.

It is much harder to count partitions, and there are, in general, few neat formulae. There are two useful techniques we can use: Ferrers diagrams (introduced in [4, p55]) and generating functions.

A Ferrers diagram represents a partition as an array of left-aligned rows of dots. For example, the diagram showing the partition 6 5 4 3 2 of 20 is shown in figure 4.1.

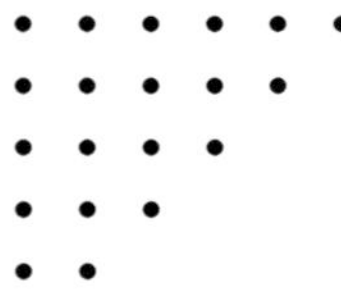

Figure 4.1

Problem 4.1

Let $f(n)$ be the number of partitions of n in which the largest part is at most 3 and let $g(n)$ be the number of partitions of n with at most 3 parts. Prove that $f(n) = g(n)$.

Consider the Ferrers diagram of a partition with largest part at most three, for example 3 3 2 1 1.

Figure 4.2

What is it about figure 4.2 which tells us that the partition satisfies the condition? In a Ferrers diagram, the lengths of the rows represent the size of the parts; it follows that the largest part is the length of the first row, which is 3. Hence the first condition is equivalent to the fact that the first row has length at most 3.

We wish to associate this with a diagram for a partition with at most 3 parts. The number of parts in a partition is the length of the first column. So the second condition is equivalent to the fact that the first column has length at most 3. This suggests a natural process: swap the rows and columns. This produces the Ferrers diagram shown in figure 4.3, which represents the partition 5 3 2.

Figure 4.3

We have created a bijection between partitions with largest part at most 3 and partitions with at most 3 parts, and it follows immediately that $f(n) = g(n)$.

The two Ferrers diagrams (and their associated partitions) are known as *conjugates*.

Problem 4.2

Let $f(n)$ be the number of partitions of n in which the largest part is odd and the smallest part is larger than half the largest. Let $g(n)$ be the number of partitions of n with unique smallest part in which the largest part is at most twice the smallest. Prove that $f(n) = g(n)$.

Taking $n = 3$ as an example, the partition 7 6 6 5 4 of 28, which is of the first type, is represented by the Ferrers diagram in figure 4.4.

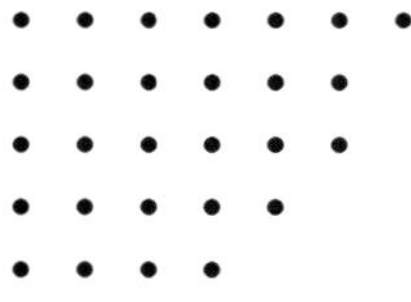

Figure 4.4

Separate this diagram into two parts. There is a 5×4 block on the left, and on the right there is a Ferrers diagram for 3 2 2 1. Place this below the block to produce the diagram on the left in figure 4.5, and then form the conjugate diagram on the right by transposing rows and columns.

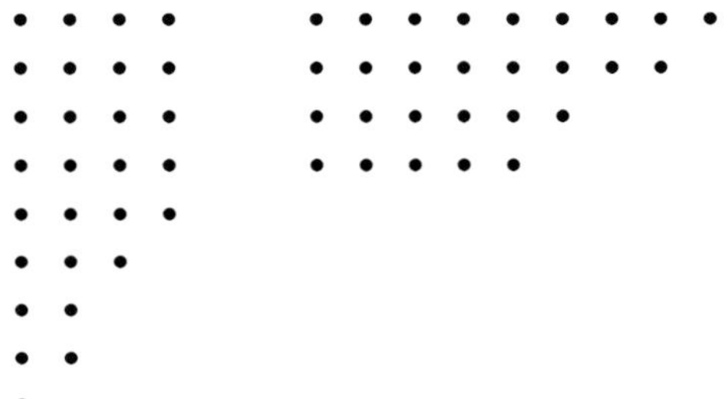

Figure 4.5

The right-hand diagram represent the partition 9 8 6 5 of 28. This satisfies the conditions for the second type of partition. Indeed, if the smallest part were not unique, a larger rectangle could have been chosen in the first step, and if the smallest part were smaller than half the largest, then the initial diagram would have the same property. Moreover, the two transformations can be reversed, so this is a bijection. It should be clear that this does not depend on the particular numbers chosen.

Generating functions were introduced as a general enumeration tool in [4, p109] and then applied to partitions. The idea is to form a power series, which looks a bit like a polynomial expression in x but has infinitely many terms. The coefficients of the series will then be interpreted as numbers of partitions. This is best illustrated by an example.

Problem 4.3

Find a generating function for partitions into positive integers.

Let us consider first how to do this for partitions of numbers up to and including 4. We must be careful that 1 3 and 3 1 are not counted as two partitions in the enumeration. The generating function turns out to be the polynomial

$$P_4(x) = (1 + x + x^2 + x^3 + x^4)(1 + x^2 + x^4)(1 + x^3)(1 + x^4)$$

It becomes clear why this works if we write the indices in the rather eccentric way

$$(1 + x^1 + x^{1+1} + x^{1+1+1} + x^{1+1+1+1})(1 + x^2 + x^{2+2})(1 + x^3)(1 + x^4).$$

When this polynomial is expanded in the usual way, by choosing powers in the four brackets, the term in x^4 arises from the five products $x^4 \times 1 \times 1 \times 1$, $x^2 \times x^2 \times 1 \times 1$, $x \times 1 \times x^3 \times 1$,$1 \times x^4 \times 1 \times 1$ and $1 \times 1 \times 1 \times x^4$, so it is $5x^4$. This corresponds to the five partitions of 4, and now we can see what is going on by pairing up products and partitions as shown in table 4.1.

Product	Partition
$x^4 \times 1 \times 1 \times 1$	1 1 1 1
$x^2 \times x^2 \times 1 \times 1$	1 1 2
$x \times 1 \times x^3 \times 1$	1 3
$1 \times x^4 \times 1 \times 1$	2 2
$1 \times 1 \times 1 \times x^4$	4

Table 4.1

The four terms in each product represent what the term taken from the corresponding bracket in the polynomial is. Thus the x^4 in the first row is actually the term $x^{1+1+1+1}$, showing that the 1 appears four times in the partition, and the three 1s indicate that neither 2, 3 or 4 are used. In the second row, the product shows that we have selected x^{1+1} from the first bracket, x^2 from the second and nothing from the third and fourth brackets. This generating function shows, in the coefficients of x, x^2 and x^3, the partitions of 1, 2 and 3, but the coefficient of x^5 does not represent the partitions of 5, since, for example, the partition 1 1 1 1 1 is omitted, since it would require a term of x^5 in the first bracket.

The generating function for all non-negative integers requires power series rather than polynomials. Writing $p(n)$ for the number of partitions of n, it is

$$\sum p(n)x^n = (1 + x + x^2 + \cdots)(1 + x^2 + x^4 + \cdots)(1 + x^3 + x^6 + \cdots)\cdots$$

which can be written in the neater form

$$\sum p(n)x^n = \frac{1}{(1-x)(1-x^2)(1-x^3)\cdots} = \frac{1}{\prod_{n\geq 1}(1-x^n)}$$

using the formula for the sum to infinity of a geometric progression. This infinite product is due to Euler.

We now use generating functions in a more challenging context.

Problem 4.4

Prove that the number of partitions of n into odd parts is equal to the number of partitions of n into distinct parts.

The generating function for the number of partitions of a positive integer into odd parts is given by the infinite product

$$O(n) = (1 + x + x^2 + x^3 + \cdots)(1 + x^3 + x^6 + x^9 + \cdots)$$
$$(1 + x^5 + x^{10} + x^{15} + \cdots)\cdots .$$

This works in a similar way to that of problem 4.3 if you write the indices as sums of odd numbers. Now we see that

$$O(n) = (1 + x^1 + x^{1+1} + x^{1+1+1} + \cdots)(1 + x^3 + x^{3+3} + x^{3+3+3} + \cdots)$$
$$(1 + x^5 + x^{5+5} + x^{5+5+5} + \cdots)\cdots$$

For example, $x^{1+1+1+3+5+5+11+11+11+17}$ will represent the partition 1 1 1 3 5 5 11 11 11 17 of 66 into odd parts. Consequently, the coefficient of x^{66} in the infinite product will be the number of ways of partitioning 66 into odd parts. As in problem 4.3, we can express this in the form

$$O(n) = \frac{1}{(1-x)(1-x^3)(1-x^5)\ldots}.$$

By a similar argument, the generating function for the number of partitions into distinct parts is

$$D(x) = (1+x)(1+x^2)(1+x^3)(1+x^4)\ldots$$

as can be seen by multiplying it out. However

$$\begin{aligned}
D(x) &= (1+x)(1+x^2)(1+x^3)(1+x^4)\ldots \\
&= \frac{(1-x)(1+x)}{(1-x)} \times \frac{(1-x^2)(1+x^2)}{(1-x^2)} \times \frac{(1-x^3)(1+x^3)}{(1-x^3)} \times \ldots \\
&= \frac{(1-x^2)(1-x^4)(1-x^6)(1-x^8)\ldots}{(1-x)(1-x^2)(1-x^3)(1-x^4)\ldots} \\
&= \frac{1}{(1-x)(1-x^3)(1-x^5)\ldots} \\
&= O(n)
\end{aligned}$$

so the two generating functions are the same, and the result follows.

In [4] this problem was set as an exercise, and a bijection argument using binary representation was offered.

4.2 A theorem of Euler

In problem 4.3 we showed that the generating function of the partition numbers was given by

$$\sum p(n)x^n = \frac{1}{\prod_{n\geq 1}(1-x^n)}.$$

We note that the denominator on the right hand side is similar to the function $D(x)$ except that the $+$ signs have been replaced by $-$ signs.

It is worth considering this product, which we will call $\bar{D}(x)$, in some detail. We have

$$\bar{D}(x) = (1-x)(1-x^2)(1-x^3)\cdots = \prod(1-x^r). \qquad (4.1)$$

When expanded, we obtain the series

$$\bar{D}(x) = 1 - x - x^2 + x^5 + x^7 - x^{12} - x^{15} + x^{22} + x^{26} - x^{35} - \ldots$$

which suggests that most of the coefficients are zero, and that the non-zero coefficients are either $+1$ or -1.

As with $D(x)$, the coefficient of x^n in $\bar{D}(x)$ is obtained by finding all the ways of selecting powers of x or 1 from each bracket of (4.1) so that the sum of the powers chosen is n. For example, we could select $-x$, $-x^2$ and $-x^4$ (and 1s otherwise), resulting in $-x^7$, or $-x$ and $-x^6$ to produce $+x^7$. The coefficient of x^7 in the final expansion is obtained by adding $+1$ or -1 for every partition of 7 into distinct parts. We obtain $+1$ whenever the partition in question has an even number of terms, and -1 when it has an odd number of terms. (For convenience, we will call partitions into an even number of distinct terms *even partitions* and define *odd partitions* similarly.)

Hence, defining $d_e(n)$ as the number of even partitions of n and $d_o(e)$ as the number of odd partitions of n, we see that $\bar{D}(x)$ is the generating function for $\bar{d}(n) = d_e(n) - d_o(n)$.

Problem 4.5

Prove that all the coefficients of $\bar{D}(x)$ are either -1, 0 or $+1$.

For small values of n we obtain table 4.2 below, which also includes the values of $d(n)$.

n	1	2	3	4	5	6	7	8	9	10	11	12
$d(n)$	1	1	2	2	3	4	5	6	8	10	12	15
$d_e(n)$	0	0	1	1	2	2	3	3	4	5	6	7
$d_o(n)$	1	1	1	1	1	2	2	3	4	5	6	8
$\bar{d}(n)$	-1	-1	0	0	1	0	1	0	0	0	0	-1

Table 4.2

Now it is clear that $\bar{d}(n)$ can take the value 0 only if $d(n)$ is even. What the values in table 4.2 seem to indicate is that this always happens for even values of $d(n)$, and thus that it might be possible to find a bijection between odd and even partitions. The values also seem to indicate that, when $d(n)$ is odd, the bijection works for most of the partitions of n into distinct parts, but that there is one awkward partition which cannot be transformed in this way.

To see how such a bijection might be constructed, we consider the eight partitions of 9 into distinct parts.

9	8 1	6 2 1
	7 2	5 3 1
	6 3	4 3 2
	5 4	

There are four even and four odd partitions, so they can certainly be paired up, but we want a pairing which is natural and general.

We might begin by pairing up the partition 9 with the partition 8 1. This immediately suggests a more general rule: a partition with smallest part 1 could be paired with the partition formed by taking that 1 and adding it to the largest part of the partition. Pairs formed in this way clearly consist of one odd and one even partition.

Applying this rule to the $n = 9$ example we obtain the pairings (8 1 $\leftrightarrow$ 9), (6 2 1 $\leftrightarrow$ 7 2) and (5 3 1 $\leftrightarrow$ 6 3). This means the final pair must be (5 4 $\leftrightarrow$ 4 3 2), but we would like to understand this pairing as an example of something more general.

So far our pairing strategy can be summarised as follows:

- If a partition has smallest part 1, then this 1 is used to make the largest part larger.
- If a partition does not have smallest part 1, then we try the reverse process: reduce the largest part by one and add a part of size 1 to the partition.

Unfortunately with the partition 5 4, neither option is available, since if we make the 5 one smaller we end up with the partition 4 4 1, which violates the condition that the parts should be distinct. We must therefore try something different.

To preserve the fact that the parts are distinct we can remove one from both the largest parts and use these two to form an additional part of size 2. This manoeuvre is similar to removing one from the largest part.

Figure 4.6

If we look at the Ferrers diagram for 5 4 on the left of figure 4.6, we see that, rather than simply moving the last dot on the longest row to the bottom, the whole of what we may call the *last diagonal* has been moved. The dots which move have been emphasised with a grey line.

If we had started with the partition 4 3 2 on the right of figure 4.6, this manoeuvre would have been impossible since the last diagonal must be smaller than the (possibly trimmed) smallest part to ensure parts remain distinct. However, we are able to reverse the process and use the smallest part of 4 3 2 to form a new last diagonal.

Now we are in a position to try and describe a general method for pairing up partitions into distinct parts. The method will always add or remove a single part, and will therefore pair up odd partitions with even ones and vice versa.

Let s be the size of the smallest part of a partition and let d be the length of the last diagonal.

- If $s > d$ we remove the last diagonal and use it to form a new part of size d.
- If $s \leq d$ we remove the smallest part and use it to make a new last diagonal of size s.

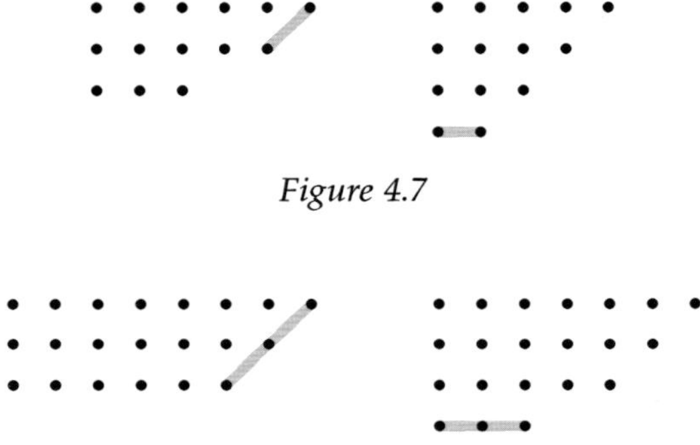

Figure 4.7

Figure 4.8

Figures 4.7 and 4.8 show this process being applied to specific partitions of 14 and 21 respectively. You should take a moment to convince yourself that if you apply the process to some partition and then apply it to the resulting partition, you get back to where you started. This means that the process really does form pairs of partitions rather than longer chains.

This is all very encouraging, but, as expected, a few partitions are *awkward* in the sense that they cannot be found partners using the scheme described.

Figure 4.9 *Figure 4.10*

Figure 4.9 shows the partition 6 5 4 of 15. We have $s = 4$ and $d = 3$, so our first thought is to remove the last diagonal and use it to form a new part of size 3 at the bottom of the Ferrers diagram. However, doing this does not produce a partition into distinct parts. This disaster occurs because two very particular conditions are met. The first is that the last diagonal overlaps with the smallest part of the original partition, so removing it makes the original smallest part smaller. The second condition is that $s = d + 1$ which means that when the original smallest part is made one smaller it is equal to the proposed new part.

The partition 5 4 3 shown in figure 4.10 is also awkward. Here $s \leq d$ so we would like to move the bottom row and use it to form a new last diagonal. However, the last diagonal overlaps with the smallest part, so removing the smallest part makes that diagonal one smaller. This caused no difficulty with the partition 5 4 3 2 shown on the right of figure 4.7, but it is a problem here precisely because $s = d$. If we remove the bottom row of the Ferrers diagram and then add a dot to the first s rows, we obtain the partition 6 5 1. This does have distinct parts, but it is not an even partition. Moreover, 6 5 1 is already paired with the partition 7 5 and cannot have multiple partners.

The fact that we have found awkward partitions of 12 and 15, and that these partitions are both odd, explains the $-x^{12}$ and $-x^{15}$ in the expansion of $\bar{D}(x)$, and the other non-zero coefficients can also be explained in this way.

Indeed, we can describe these awkward partitions precisely, since we know that the bottom row is of length s and there are d rows increasing by 1 each time as we go up the Ferrers diagram.

- If $s = d + 1$ and the last diagonal meets the bottom row, then we have an awkward partition into d parts of the number

  $$n = (d+1) + (d+2) + \cdots + (d+d) = \tfrac{1}{2}d(3d+1).$$

- If $s = d$ and the last diagonal meets the bottom row, then we have an awkward partition into d parts of the number

  $$n = d + (d+1) + (d+2) + \cdots + (d+d-1) = \tfrac{1}{2}d(3d-1).$$

In both cases we have an unpaired partition into an odd or an even number of parts according to the parity of d.

This result, which is due to Euler, is an example of proving an algebraic result by combinatorial means. It can be written, using an infinite series where k is allowed to range over both positive and negative integers, as

$$\bar{D}(x) = \sum_{k=-\infty}^{\infty} (-1)^k x^{\frac{1}{2}k(3k-1)}. \tag{4.2}$$

In exercise 4a you are asked to justify this way of writing the expansion.

4.3 Generalised pentagonal numbers

Problem 4.6

Is there any significance to numbers of the form $\frac{1}{2}d(3d \pm 1)$ for $d \geq 1$?

The sequences 1, 5, 12, 22, 35,... and 2, 7, 15, 26, 40,... generated by the two types of awkward partition turn out to be examples of *figurate numbers*.

Figurate numbers are those which can be represented as dots forming a geometrical figure. The triangular numbers $\frac{1}{2}r(r+1)$ and the square numbers r^2 are the simplest examples, and the resulting patterns look like the outlines of triangles and squares. Next we have the pentagonal numbers, which are illustrated in figure 4.11.

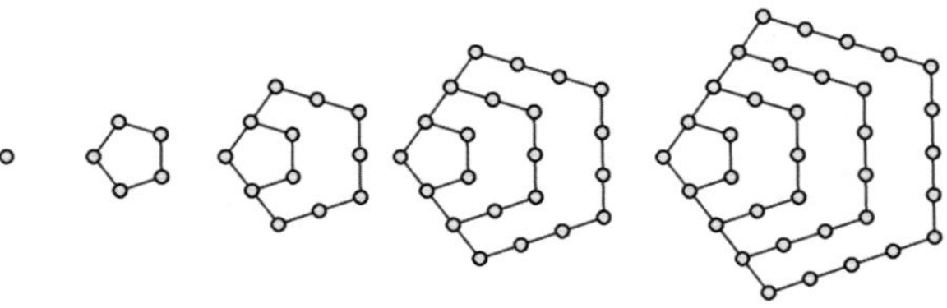

Figure 4.11

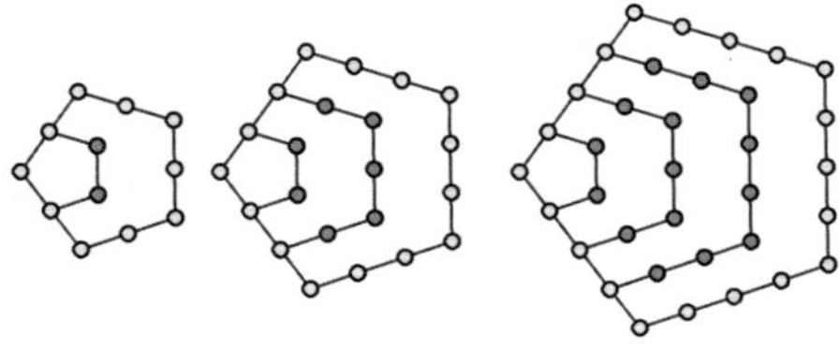

Figure 4.12

Counting all the dots shows that the first five pentagonal numbers are 1, 5, 12, 22 and 35. If we only count the dots inside the outermost pentagons, shaded grey in figure 4.12, we obtain the sequence 2, 7, 15, 26,.... The formulae for these two sequence are precisely those that give the sizes of the two types of awkward partitions, namely $\frac{1}{2}r(3r-1)$ and $\frac{1}{2}r(3r+1)$ respectively. You are invited to prove this claim in exercise 4a.

The combined sequence, given by $\frac{1}{2}r(3r \pm 1)$, is known as the sequence of *generalised pentagonal numbers*. As a result of this identification, (4.2) is known as *Euler's pentagonal number theorem*.

Exercise 4a

1. Let $p_k(n)$ be the number of partitions of n into k parts. Prove that $p_k(n+k) = \sum_{i=1}^{k} p_i(n)$.

2. Let $q_k(n)$ be the number of partitions of n into k distinct parts. Prove that $q_k(n + \frac{1}{2}k(k+1)) = \sum_{i=1}^{k} p_i(n)$.

3. Let $p^k(n)$ (for $1 \leq k \leq n$) be the number of partitions of n which do not involve the part k. Prove that $p^k(n) = p(n) - p(n-k)$.

4. Prove that the number of partitions of n with no part repeated more than k times is equal to the number of partitions of n with no part divisible by $k+1$.

5. Prove the formula $\frac{1}{2}r(3r-1)$ for the rth pentagonal number.

6. Prove the formula $\frac{1}{2}r(3r+1)$ for the 'other' pentagonal numbers.

7. Check that the use of negative powers in formula (4.2) takes care of both kinds of pentagonal number.

4.4 Partitioning sets

We first consider problems about dividing a set into subsets.

Problem 4.7

In how many ways can the set $\{1,2,3,4,5,6\}$ be partitioned into disjoint non-empty subsets?

We begin with some simple observations, but it is as well to be clear about them.

Note first that sets cannot contain repeated elements and the order of objects in a set does not matter. Hence the lists $\{1,2,4\}, \{1,2,1,4\}$ and $\{4,1,2\}$ all refer to the same set. We might as well place the elements in numerical order, so the usual description of this set would be $\{1,2,4\}$.

In the problem we are addressing, note that the order of the subsets does not matter. In other words, the two partitions $\{1\},\{2,5\},\{3,4,6\}$ and $\{2,5\},\{3,4,6\},\{1\}$ are considered to be the same. However, we count the partitions $\{1\},\{2,5\},\{3,4,6\}$ and $\{1\},\{2,6\},\{3,4,5\}$ as different, since they involve a different trio of subsets.

More importantly, note that the problem does not specify how many subsets there are and how big they are. It is important that none of the subsets is empty, but apart from that we are not specifying how the six elements are going to be split up. At one extreme we will have the partition into a single set, namely $\{1,2,3,4,5,6\}$, and at the other extreme we could split the set into six 'singleton' sets, namely $\{1\},\{2\},\{3\},\{4\},\{5\},\{6\}$.

It is instructive to consider a different problem: in how many ways can the set $\{1,2,3,4,5,6\}$ be partitioned into a subset of size 1, a subset of size 2 and a subset of size 3? In fact, this is a much simpler problem. It is enough to call these three subsets A, B and C, in the order specified. We can now describe a partition of $\{1,2,3,4,5,6\}$ by allocating letters to the six elements in order. For example, the partition $\{1\},\{2,6\},\{3,4,5\}$ would be given the 'code' $ABCCCB$. The number of such partitions would then be the number of such codes. But this is just the number of anagrams of the word $ABBCCC$. This is the multinomial coefficient $\binom{6}{1\,2\,3} = \frac{6!}{1!\times 2!\times 3!}$.

Problem 4.7 is significantly harder than this, although it is possible to tackle it using multinomial coefficients. We split the problem into 11 cases according to the different partitions of the number 6. Each of these gives rise to a multinomial coefficient; for example, the partition $1+1+1+3$ will correspond to $\binom{6}{1\,1\,1\,3}$. Summing over all partitions of 6, we obtain the correct numerical solution.

However, this would be impossible for much larger sets, such as $\{1, 2, \ldots, 100\}$ (and in fact the method is impractical even for smaller sets since, for instance, $p(10) = 42$). Hence we shall avoid this cumbersome method and look for something better.

We shall consider the general case about splitting the set $\{1, 2, 3, \ldots, n\}$ into subsets so that none is empty and no two intersect. We denote the number of ways of doing this by B_n, and note we are asked for the value of B_6.

As always, it is sensible to look at small cases. The results for $n = 1, 2$ and 3 are given in table 4.3.

n	Partitions	B_n
1	$\{1\}$	1
2	$\{1\} \cup \{2\}$ $\{1,2\}$	2
3	$\{1\} \cup \{2\} \cup \{3\}$ $\{1,2\} \cup \{3\}$ $\{1\} \cup \{2,3\}$ $\{1,3\} \cup \{2\}$ $\{1,2,3\}$	5

Table 4.3

It will turn out to be useful to define $B_0 = 1$, which makes sense since there is nothing you can do to the empty set $\varnothing$ apart from leave it alone. Some of you might recognise these as the first four Catalan numbers, C_n — see, for example, [4, p80]. However, it is unwise to jump to conclusions, so we look at the next term in the sequence.

We need to partition $\{1, 2, 3, 4\}$ into non-empty subsets. Clearly the number 4 is special because it is new. It could be put into a set on its own, or it could be combined with any of the previous sets. The first alternative is equivalent to combining it with the empty set. The number of ways of doing this is B_3, since it counts the five partitions

$$\{1\} \cup \{2\} \cup \{3\} \cup \{4\} \quad \{1,2\} \cup \{3\} \cup \{4\} \quad \{1\} \cup \{2,3\} \cup \{4\}$$
$$\{1,3\} \cup \{2\} \cup \{4\} \quad \{1,2,3\} \cup \{4\}.$$

Alternatively, we could place the 4 in a non-empty set in the list for B_3. This creates ten new partitions, namely

$$\{1,4\}\cup\{2\}\cup\{3\} \quad \{1\}\cup\{2,4\}\cup\{3\} \quad \{1\}\cup\{2\}\cup\{3,4\} \quad \{1,2,4\}\cup\{3\}$$
$$\{1,2\}\cup\{3,4\} \quad \{1,4\}\cup\{2,3\} \quad \{1\}\cup\{2,3,4\}$$
$$\{1,3,4\}\cup\{2\} \quad \{1,3\}\cup\{2,4\} \quad \{1,2,3,4\}$$

so we have $B_4 = 15$. This shows that we are not dealing with the Catalan sequence, since $C_4 = 14$.

We have shown that, in order to deduce the value B_4 from that of B_3, we needed to know the structure of the partitions of $\{1,2,3\}$. Acting on this insight, let us reorganise the information we already have. We call the sets to which the new number 4 is added *seed sets,* and arrange the partitions based on the size of these sets as shown in table 4.4. This is useful, since it allows us to amalgamate the two processes of adding to an empty set and a non-empty set into one.

Size of seed set	Partitions of $\{1,2,3,4\}$	Number
0	$\{1\}\cup\{2\}\cup\{3\}\cup\{4\}\{1,2\}\cup\{3\}\cup\{4\}$ $\{1\}\cup\{2,3\}\cup\{4\}\{1,3\}\cup\{2\}\cup\{4\}$ $\{1,2,3\}\cup\{4\}$	5
1	$\{1,4\}\cup\{2,3\}$ $\{1,4\}\cup\{2\}\cup\{3\}$ $\{2,4\}\cup\{1,3\}$ $\{2,4\}\cup\{1,3\}$ $\{2,4\}\cup\{1\}\cup\{3\}$ $\{3,4\}\cup\{2,3\}$ $\{3,4\}\cup\{1\}\cup\{2\}$	6
2	$\{1,2,4\}\cup\{3\}$ $\{1,3,4\}\cup\{2\}$ $\{2,3,4\}\cup\{1\}$	3
3	$\{1,2,3,4\}$	1

Table 4.4

As before we have $B_4 = 15$. Now let us try and represent the numbers in the right-hand column in terms of the previous terms of the sequence. The first is B_3 for the reason already stated.

When the 4 is added to a set of size 1, one element is chosen from $\{1,2,3\}$ and the other two are partitioned. There are $\binom{3}{1}$ ways of choosing the element, and B_2 ways of partitioning the remaining two, so the number of new partitions in this row of the table is $\binom{3}{1}B_2$, which is 6. When the 4 is added to a set of size 2, two elements are chosen from $\{1,2,3\}$ and the other one is partitioned. There are $\binom{3}{2}$ ways of choosing the pair and B_1 ways of partitioning the remaining one, so the number of new partitions in this row of the table is $\binom{3}{2}B_1$, which is 3. Finally, when the 4 is added to a set of size 3, there is only one way of choosing this set and nothing else has to be done, so there is 1 (or B_0) ways of doing this. We therefore have (after including two more binomial coefficients which are equal to 1)

$$B_4 = \binom{3}{0}B_3 + \binom{3}{1}B_2 + \binom{3}{2}B_1 + \binom{3}{3}B_0.$$

This argument can be generalised. Assume that we know the values of $B_1, B_2, \ldots, B_n$ and think about partitioning the set $\{1,2,\ldots,n+1\}$. The new number $n+1$ must be put into one of the sets into which $\{1,2,3,\ldots,n\}$ has been partitioned or into the empty set. Suppose that this has size k, where $0 \leq k \leq n$. There are $n-k$ remaining elements, $\binom{n}{n-k}$ ways of choosing them and B_{n-k} ways of partitioning them. Hence the number of such partitions is $\binom{n}{n-k}B_{n-k}$. Hence

$$B_{n+1} = \binom{n}{n}B_n + \binom{n}{n-1}B_{n-1} + \binom{n}{n-2}B_{n-2} + \cdots + \binom{n}{0}B_0.$$

Writing this using summation notation, we have the formula

$$B_{n+1} = \sum_{k=0}^{n} \binom{n}{k} B_k.$$

Continuing this process, we have

$$\begin{aligned}
B_0 &= B_1 = 1\\
B_2 &= 1 \times B_1 + 1 \times B_0 = 2\\
B_3 &= 1 \times B_2 + 2 \times B_1 + 1 \times B_2 = 5\\
B_4 &= 1 \times B_3 + 3 \times B_2 + 3 \times B_1 + 1 \times B_0 = 15\\
B_5 &= 1 \times B_4 + 4 \times B_3 + 6 \times B_2 + 4 \times B_1 + 1 \times B_0 = 52\\
B_6 &= 1 \times B_5 + 5 \times B_4 + 10 \times B_3 + 10 \times B_2 + 5 \times B_1 + 1 \times B_0 = 203
\end{aligned}$$

and now we have solved problem 4.7.

The numbers B_n are called Bell numbers. They are named after Eric Temple Bell, most famous as the author of *Men of Mathematics* but also a mathematician in his own right. As so often, the name is a misattribution, since the numbers were studied decades before he was born. In fact, these set partitions made an appearance in a problem called *The Tale of Genji* from mediaeval Japan.

Problem 4.8

In how many ways can the set $\{1,2,\ldots,n\}$ be partitioned into exactly r disjoint non-empty subsets, where $1 \leq r \leq n$?

Again the order of the subsets is immaterial. We shall call this number $S(n,r)$.

It is obvious that $B_n = \sum_{r=1}^{n} S(n,r)$, but it is not clear how to split the Bell numbers up to produce the numbers we require. However, we know that $S(n,1) = S(n,n) = 1$, and that $S(n,r) = 0$ when $r > n$. We now consider what happens to the number $n+1$ when we partition $\{1,2,\ldots,n+1\}$ into r subsets for $r \geq 2$.

Either it is placed into a subset on its own, in which case the set $\{1,2,\ldots,n\}$ is partitioned into $r-1$ subsets, and this can done in exactly $S(n,r-1)$ ways. Or it is placed in one of the r sets in which $\{1,2,\ldots,n\}$ is partitioned, and this can be done in $rS(n,r)$ ways. We now have a recurrence relation

$$S(n+1,r) = S(n,r-1) + rS(n,r) \qquad \text{for } 2 \leq r \leq n. \tag{4.3}$$

The numbers can be generated from the array shown in table 4.5. Each $S(n,r)$ appears in the rth column of the nth row, and the numbering of the rows and columns begins with 1.

1					
1	1				
1	3	1			
1	7	6	1		
1	15	25	10	1	
1	31	90	65	15	1

Table 4.5

The entries in the sixth row (apart from the initial and final 1s) are determined as follows:

$$\begin{aligned} 1 + 2 \times 15 &= 31 \\ 15 + 3 \times 25 &= 90 \\ 25 + 4 \times 10 &= 65 \\ 10 + 5 \times 1 &= 15 \end{aligned}$$

and the Bell numbers can be recovered by summing the entries in each row, giving the values 1, 2, 5, 15, 52 and 203 as before.

The numbers in this array are called *Stirling numbers of the second kind*. They are named after the Scottish mathematician James Stirling who also found an approximate formula for $n!$. We will meet Stirling numbers of the first kind in section 4.5.

Exercise 4b

1. Prove that

(a) $S(n,2) = 2^{n-1} - 1$;

(b) $S(n,3) = \frac{1}{2}(3^{n-1} - 2^n - 1)$;

(c) $S(n,n-1) = \frac{1}{2}n(n+1)$;

(d) $S(n,k) = \frac{1}{k!}\sum\binom{n}{n_1\, n_2 \ldots n_k}$.

where the sum is taken over all solutions in positive integers of the equation $n_1 + n_2 + \cdots + n_k = n$.

2. Prove that $S(n,k) = \sum_{r=0}^{n-1}\binom{n-1}{r}S(n-r-1,k-1)$.

3. Prove that, if $N \geq n$, $\sum_{r=1}^{n} r!\binom{N}{r}S(n,r) = N^n$.

4. An array of integers is defined by the recurrence relation:

$$u(n,0) = a_n \qquad (n \geq 0);$$
$$u(n+1,r+1) = u(n,r) + u(n+1,r) \qquad (0 \leq r \leq n-1).$$

Here n denotes the row number and r the column number. Note that both rows and columns are numbered from 0, and that there are $n+1$ entries in the nth row.

(a) If $a_k = 1$ for all $k \geq 0$, identify the terms in the array.

(b) Find a formula for $u(n,r)$ in terms of $a_0, a_1, \ldots, a_n$.

(c) If $a_r = B_r$ for $0 \leq r \leq n$, identify the term $u(n,n)$.

(d) Hence devise a way of ensuring that $u(n,n) = B_n$. (The resulting array is known as the Bell table.)

(e) Find a combinatorial interpretation of the entry $u(n,r)$ in the Bell table.

5. Suppose that N is the product of n different primes. Prove that B_n is the number of ways of writing N as a product of integers greater than 1 (in which the order of factors does not matter).

6. A *rhyme-scheme* describes the pattern of rhyme between the lines of a poem. If each line of the poem is replaced by a letter, so that lines which rhyme have the same letter and lines which do not rhyme have different letters, the rhyme-scheme is the pattern of letters. For instance, the ditty

A graduate student at Trinity
Computed the square of infinity.
But it gave him the fidgets
To put down the digits
So he dropped math and took up divinity.

has the rhyme-scheme *aabba*, whereas

A mathematician called Lee
Was stung on the nose by a wasp.
When asked if it hurt

He said: No, not at all!
It can do it again if it wants.

has the scheme *abcde*.
Prove that B_n counts the number of rhyme-schemes of a poem with n lines.

4.5 Partitioning permutations

Another sort of partitioning occurs when permutations are expressed as a product of disjoint cycles.

For example, the permutation $\begin{pmatrix} 1 & 2 & 3 & 4 & 5 & 6 & 7 & 8 \\ 4 & 2 & 8 & 7 & 6 & 5 & 1 & 3 \end{pmatrix}$ can be expressed as $(1\,4\,7)(2)(3\,8)(5\,6)$. It is normal to omit the 1-cycle (2) but we will be including it.

Problem 4.9

How many permutations of n elements consist of r disjoint cycles?

We shall denote this number by $s(n,r)$. Let us examine this problem when $n = 4$.

If $r = 1$ then the permutation consists of a single 4-cycle. Without loss of generality, we begin with the element 1. There are three choices for the next element and two for the third, so $s(4,1) = 6$; the permutations are

$$(1\,2\,3\,4) \quad (1\,2\,4\,3) \quad (1\,3\,2\,4) \quad (1\,3\,4\,2) \quad (1\,4\,2\,3) \quad (1\,4\,3\,2).$$

When $r = 2$ there are two kinds of permutation. There are eight which consist of a 3-cycle and a 1- cycle (since there are four ways of choosing the triple of elements and two ways of ordering them) and three products of two 2-cycles (since 1 can be paired with either 2, 3 or 4). Hence $s(4,2) = 11$.

When $r = 3$, there must be a 2-cycle and two 1-cycles, and there are six of these, so $s(4,3) = 6$. Finally, when $r = 4$, there is only one permutation, the identity. Hence $s(4,4) = 1$.

Adding, we obtain all 24 permutations with four elements.

It would seem, on the basis of this investigation, that the problem breaks down into three parts. First we have to enumerate the partitions of 4. In order, these are 4, 31, 2 2, 211 and 1111 . Now we find the number of ways of allocating the elements to the resulting cycles. Finally, we regroup according to the number of cycles; in this case this meant amalgamating the results for the partitions 31 and 2 2 . Now it is fairly clear that this is not a feasible strategy in general; we have already seen how difficult it is to enumerate partitions. Perhaps, then, it is possible to create a recurrence relation?

Suppose we have a permutation of $\{1,2,\ldots,n+1\}$ with $r+1$ cycles. We try to relate this to a permutation of $\{1,2,\ldots,n\}$. It seems sensible to consider what happens to the 'new' element $n+1$.

If it is a fixed point of the permutation, then the other elements constitute a permutation of $\{1,2,\ldots,n\}$ with r cycles, and we know that there are $s(n,r)$ of these. If not, then it is sent to another element; there are n possibilities for this. Consider the cycle in which $n+1$ occurs and remove this number, producing a permutation of $\{1,2,\cdots,n\}$ with $r+1$ cycles. There are $s(n,r+1)$ of these. Note that this process is reversible; we can select any permutation of $\{1,2,\ldots,n\}$ and insert a new element $n+1$ before an arbitrary element to produce a permutation of $\{1,2,\ldots,n+1\}$, or we can just add a new 1-cycle consisting of $n+1$ alone. Hence we have a bijection between permutations of $\{1,2,\ldots,n\}$ and permutations of $\{1,2,\ldots,n+1\}$. As a result, we have the recurrence relation

$$s(n+1,r+1) = s(n,r) + ns(n,r+1)$$

with the initial conditions $s(n,1) = (n-1)!$ and $s(n,n+1) = 0$

The $s(n,r)$ are known as *Stirling numbers of the first kind.* They are shown in table 4.6, where the numbering of the rows and columns begins with 1.

1					
1	1				
2	3	1			
6	11	6	1		
24	50	35	10	1	
120	274	225	85	15	1

Table 4.6

A natural question to ask is how the two kinds of Stirling numbers are related. Both of them relate to a set $\{1,2,\ldots,n\}$ which is split into r non-empty disjoint subsets. For Stirling numbers of the second kind, that is it; we simply count the ways of doing this. For Stirling numbers of the first kind, there is a further step, in that each set is then arranged as a cycle. This is the same thing as seating the elements of the set around a circular table, so for a subset containing r elements, there are $(r-1)!$ ways of doing this.

A different approach to evaluating Stirling numbers of the second kind is to fix the value of k, the number of parts, and to consider the numbers $\{1,2,\ldots,n\}$ which are to be partitioned.

We will illustrate this in the case when $n = 10$ and $k = 4$. Consider, as an example, the partition

$$\{3,5,10\},\{9\},\{1,2,4,8\},\{6,7\}.$$

We allocate labels to the sets. The set containing 1 is labelled A. The number 2 also belongs to A, but 3 is in a new set, which is labelled B. The numbers 4 and 5 belong to sets we have already labelled, but 6 belongs to a new set which we call C. The numbers 7 and 8 are already allocated, but 9 is the sole member of a new set which we label D. Finally 10 is in set B and we obtain the full labelling shown in table 4.7.

1	2	3	4	5	6	7	8	9	10
A	A	B	A	B	C	C	A	D	B

Table 4.7

Now suppose that all we know is the first instance of each subset name, so we have table 4.8.

1	2	3	4	5	6	7	8	9	10
A		B			C			D	

Table 4.8

Note that this is one of the $\binom{9}{3} = 84$ possible patterns, since B, C and D are placed in three of a possible 9 positions. How many ways are there to fill in the missing letters? The cell following A must contain another

A. The two cells following B can contain either A or B , so there are 4 ways of doing this. Similarly there are 9 choices for the cells between C and D and 4 choices for the final cell. Hence there are 144 partitions which correspond to the pattern shown. It is theoretically possible to list all the patterns and carry out this calculation for each of them, showing by summation that $S(10,4) = 34105$. In practice this method is untenable, but, as you will see in exercise 4c, it has an interesting consequence.

Exercise 4c

1. Prove that

   (a) $s(n,1) = (n-1)!$;

   (b) $s(n,n-1) = \binom{n}{2}$;

   (c) $s(n,n-2) = \frac{1}{4}(3n-1)\binom{n}{3}$;

   (d) $s(n,n-3) = \binom{n}{2} \times \binom{n}{4}$;

   (e) $s(n,2) = (n-1)\sum_{r=1}^{n-1}\frac{1}{r}$.

2. Show that a generating function for Stirling numbers of the second kind $S(n,k)$ for a fixed k is given by $G_k(x) = \frac{x^k}{\prod_{r=1}^{k}(1-rx)}$.

3. *The twelvefold way: a review of counting procedures*

   Suppose you have n balls which are to be put into m boxes such that each ball goes into exactly one box. It is natural to ask how many ways this can be done. However, we must be careful to specify the problem precisely, and there are many different ways to do this.

   First we must consider how many balls a box can contain. This could be unrestricted so that, for example, all the balls can be placed in the first box. Alternatively we could insist that each box contains at most one ball, or indeed that each box contains at least one ball. These latter options imply $n \leq m$ and $n \geq m$ respectively.

Next we ask whether the balls are distinguishable (which we will abbreviate to D) or indistinguishable (I).

Finally we ask whether or not the boxes are distinguishable. If they are we can imagine them in a row, perhaps numbered from 1 to m. If, on the other hand, we consider the boxes to be indistinguishable we need to be a little more fanciful. Rather than imagine them in a row, we might imagine the identical boxes being closed and placed in a large sack which is then vigorously shaken. Two arrangements are only counted as different if the sacks of boxes can be told apart.

These considerations lead us to twelve different interpretations of the question: *In how many ways can n balls be placed in m boxes?*

The different interpretations are shown in table 4.9 while table 4.10 shows the answers to these twelve questions in a random order. (Note that $p_m(n)$ is the number of partitions of n into m non-zero parts.)

	D boxes D balls	I boxes D balls	D boxes I balls	I boxes I balls
Unrestricted				
At most one ball per box				
At least one ball per box				

Table 4.9

Label	Formula	Label	Formula
A	1	B	m^n
C	$\frac{m!}{(m-n)!}$	D	$\binom{m}{n}$
E	$\left(\!\binom{m}{n}\!\right)$	F	$\left(\!\binom{m}{m-n}\!\right)$
G	$S(n,m)$	H	$m!S(n,m)$
I	$\sum_{i=1}^{m} S(n,i)$	J	$p_m(n)$
K	$\sum_{i=1}^{m} P_i(n)$		

Table 4.10

Use table 4.10 to complete table 4.9.

Chapter 5

Combinatorial Games

In this chapter we will discuss various games played by two fictional characters called Alice and Bob. We will be interested in whether Alice or Bob has a strategy which guarantees them victory, no matter how their opponent chooses to play. This desire for certainty means that we will not discuss any games which involve dice, cards or other random elements, since even an expert can lose a game of chance by being sufficiently unlucky. We will also ignore games where the two players effectively make decisions simultaneously. Such games, which include rock-paper-scissors and the prisoner's dilemma, possess a rich theory but do not form part of strictly combinatorial game theory.

Fortunately, although many games do not possess winning strategies of the type we would like, those that do are still varied and interesting.

5.1 Winning positions

Problem 5.1

There is a pile of twenty-one counters on a table. Alice and Bob take it in turns to remove one, two or three counters from the pile. The person who removes the last counter is the winner. If Alice goes first, who has a winning strategy?

Let us start by following a classic piece of problem-solving advice and considering a simpler problem. We can do this either by restricting the moves available or by reducing the number of counters on the table initially. The latter option preserves more of the original game's structure, so we will pursue that first.

With one, two or three counters on the table, Alice can win on her first turn by taking them all. This might seem too obvious to be worth stating. However, we should never be too proud to consider the simplest cases of a problem: they often yield insight and seldom require a great deal of effort.

Next we consider the game with four counters. Here it is obvious that Bob will win. There will be one, two or three counters on the table after Alice's turn, and Bob can simply take them all. With five counters Alice will lose if she takes two or three of them. However, if she takes only one, then Bob will be faced with a pile of four counters on his turn, and we saw earlier that he is sure to lose from this position.

Alice can also win if the initial pile contains six or seven counters, since she can ensure Bob is faced with a pile of four.

With eight counters the situation changes once again. After Alice's first turn, there will be either five, six or seven counters left. Bob can then ensure that Alice is confronted with a pile of four counters at the start of her second turn. This is bad news for Alice, so if the game starts with eight counters, then Bob has a winning strategy.

The fact that the person confronted with eight counters loses means that Alice has a winning strategy if there are nine, ten or eleven counters at the start. This in turn shows that Bob has a winning strategy if there are twelve counters and so on.

After all this work we can provide a solution to the original problem. With twenty-one counters on the table Alice has a winning strategy: she starts by taking one counter and leaving Bob with twenty. Whatever Bob does, Alice can ensure that the number of counters on the table at the end of her subsequent turns is sixteen, then twelve, then eight, then four and finally zero.

In general, if you ensure that your opponent always starts their turn facing a number of counters which is divisible by four, then you are bound to win. Thus, if the game starts with n counters on the table, Alice has a winning strategy precisely when n is not divisible by four.

Now let us introduce some terms to simplify similar discussions. We will call the states that a game can be in its *positions*. A position where the person to play will *win* if they play perfectly is called a *$\mathcal{W}$-position*, and a position where the person to play will *lose* if their opponent plays perfectly is called an *$\mathcal{L}$-position*.

These terms take a bit of getting used to, so it is worth seeing how they could have been used earlier.

In problem 5.1 the $\mathcal{L}$-positions are precisely those where the number of counters on the table is divisible by four. (We note that zero is divisible by four, which is good since the player who starts their turn facing zero counters has just lost.) All other positions are $\mathcal{W}$-positions.

The key facts which allowed us to analyse the game can be summarised as follows:

Position labelling rules

(i) The positions where the previous player has already won are $\mathcal{L}$-positions.

(ii) If it is possible to get from a given position to an $\mathcal{L}$-position, then that position is a $\mathcal{W}$-position.

(iii) If all the moves from a given position lead to $\mathcal{W}$-positions, then that position is an $\mathcal{L}$-position.

We will refer to these three rules as the *labelling rules*.

With this in place we see that if Alice goes first in any game, she has a winning strategy if the initial position is a $\mathcal{W}$-position. She simply moves to an $\mathcal{L}$-position, which must be possible by (ii). Her strategy is simply to move to an $\mathcal{L}$-position whenever she can. On Bob's turn he must move to a $\mathcal{W}$-position by (iii), so Alice can move back to an $\mathcal{L}$-position. Thus Bob can never end his turn on an $\mathcal{L}$-position so, by (i), he cannot win. Similarly, if the initial position is an $\mathcal{L}$-position, then Bob has a winning strategy.

5.2 Game graphs

We now introduce an idea which will allow us to analyse a very wide variety of games in a systematic way.

A game consists of a set of positions $\{a, b, c, \ldots\}$. If it is possible to move from position a to position b, then we say that b is a *follower* of a. All that really matters in a game is knowing which positions follow which. It

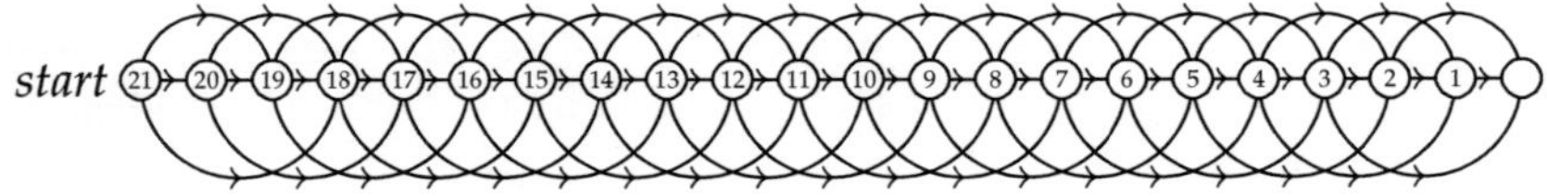

Figure 5.1

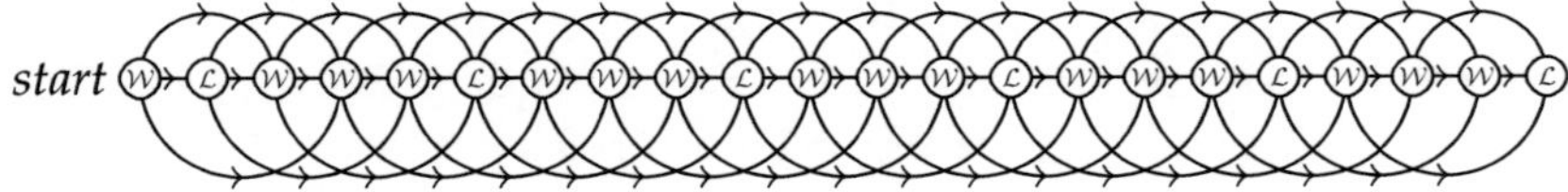

Figure 5.2

makes no difference whether the positions are piles of counters, drawings on a piece of paper or arrangements of pieces on a board. To distil out the essential relationships we can take a game and form a directed graph whose vertices are the positions and where an arrow joins a to b if b is a follower of a. This directed graph (see page 42 in chapter 2) is called the *game graph*.

For example, figure 5.1 shows the game graph for problem 5.1 and figure 5.2 shows the same graph with the vertices labelled according to whether they are $\mathcal{W}$ or $\mathcal{L}$ positions.

Our definition of a game graph includes the hidden assumption that the moves available from a given position do not depend on whose turn it is. Games where this assumption holds are called *impartial games* and most of our examples will belong to this category.

Many games played in real life are not impartial. In chess, for example, the first player can only move the white pieces while the second player can only move the black pieces. Games where the two players have different moves available to them are called *partisan games* and, while the idea of a game graph can be adapted to study such games, we will avoid them for now.

The next game we will look at was invented by one of the early pioneers of combinatorial game theory.

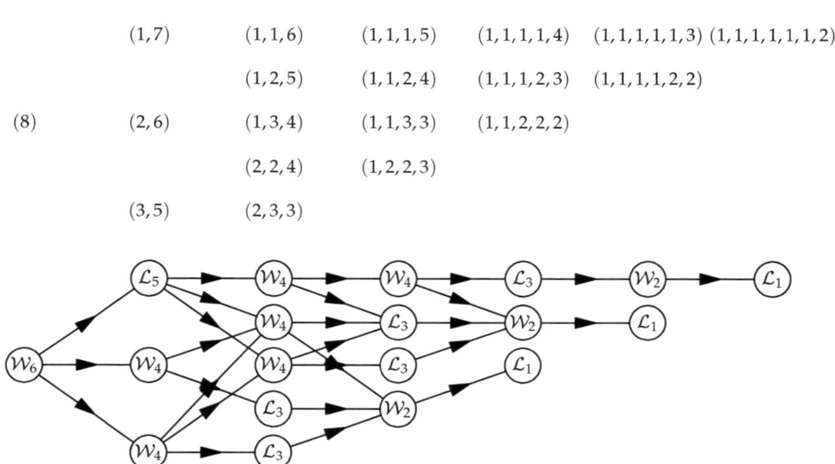

Figure 5.3

Problem 5.2

A position in *Grundy's game* consists of a number of piles of counters. Alice and Bob take it in turns to choose a pile and divide it into two *unequal* piles. The player who is unable to move loses.

If the game starts with a single pile of eight counters and Alice goes first, who has a winning strategy?

We will represent a position in the game by a list of the sizes of the piles. For example, the starting position is represented by (8), and the followers of this position are $(1,7)$, $(2,6)$ and $(3,5)$. (It is illegal to divide a pile of eight counters into two piles of four.)

The possible positions and the game graph are shown in figure 5.3. The subscripts on the $\mathcal{W}$ and $\mathcal{L}$ labels indicate a possible order in which the labels could be added.

We see that the starting position is a $\mathcal{W}$-position, so Alice has a winning strategy. She should begin by moving to $(1,7)$ because it is an $\mathcal{L}$-position. Bob will then split the pile of seven counters, but however he does this, Alice will be able to use her turn to move to the position $(1,1,2,4)$. From here Bob will go to $(1,1,1,2,3)$ allowing Alice to win by going to $(1,1,1,1,2,2)$.

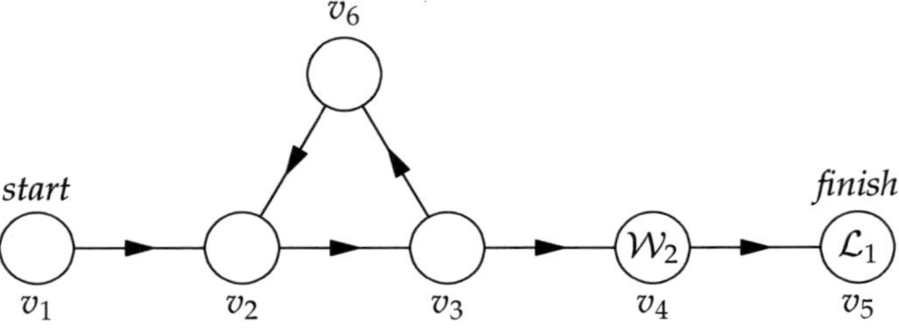

Figure 5.4

There is a potential problem with studying games via their game graphs which we must address. What if the labelling rules do not allow us to determine unambiguously whether the starting vertex is a $\mathcal{W}$ or an $\mathcal{L}$-position? The directed graph shown in figure 5.4 illustrates that this worry is justified.

We start by labelling vertex v_5 with an $\mathcal{L}$ since it is a finishing position. This implies that v_4 is a $\mathcal{W}$-position since it has an $\mathcal{L}$-position among its followers. Now, however, we have a problem. We have labelled all the positions which lead directly to $\mathcal{L}$-positions. If we are to continue, we must find a new $\mathcal{L}$-position, that is, one all of whose followers are already labelled $\mathcal{W}$. Unfortunately, all of the unlabelled vertices have unlabelled followers. We are at an impasse, and the whole idea of using game graphs seems about to collapse. What exactly has gone wrong?

If we imagine Alice and Bob actually moving a counter on this graph, the problem becomes evident. Alice moves the counter from v_1 to v_2 and Bob then moves it to v_3. Now Alice has a choice, but if she moves to v_4 she will certainly lose, so the only reasonable thing to do is to move to v_6. Now Bob moves to v_2 and Alice moves to v_3. At this point Bob has the same choice as Alice did a few moves ago. Moving to v_4 is fatal, so he will also move to v_6, and the cycle will begin again. If both players are trying to avoid losing, then they will end up in an infinite sequence of moves, and the game will never end.

These observations mean that our inability to determine whether v_1 was $\mathcal{W}$ or $\mathcal{L}$ ceases to be so disturbing. In fact, it is reassuring that the labelling rules do not succeed in labelling v_1. After all, neither player has a winning strategy in this game.

Infinite sequences of moves can prevent either player from having a winning strategy. Fortunately the next problem shows that they are, in a sense, the only obstacle to determining who wins an impartial game.

Problem 5.3

Prove that if an impartial game has a finite number of positions and no loops, then the labelling rules are able to determine whether any given position in the game is a $\mathcal{W}$ or an $\mathcal{L}$-position.

The argument is fairly straightforward. We choose a vertex which has not yet been labelled, and if it has an unlabelled follower we move to that vertex and repeat the process. This cannot go on indefinitely since the game has no loops, so we are sure to find a vertex all of whose followers are already labelled (possibly because it has no followers). If this vertex has an $\mathcal{L}$ among its followers we label it $\mathcal{W}$ and otherwise we label it $\mathcal{L}$. Either way, as long as there are unlabelled vertices in the graph we can always label one more. The graph is finite so we can eventually label every vertex.

Often it is completely obvious that a game cannot go on forever, but occasionally checking this condition requires a little thought.

Problem 5.4

A knight is placed on the corner cell of a standard 8×8 chessboard, and coordinates are chosen such that the centre of that cell is the origin and the centre of the cell in the opposite corner is at $(7,7)$. The knight is restricted to making moves described by the vectors $\binom{2}{\pm 1}$ and $\binom{\pm 1}{2}$ as shown in figure 5.5. Alice and Bob take it in turns to move the knight, and the player who is unable to make a move loses. If Alice goes first, who has a winning strategy?

At first glance it may not be totally clear that this game is sure to end after finitely many moves. To check that it does, we observe that the sum of the knight's x and y coordinates increases by at least one each turn, and can never exceed fourteen.

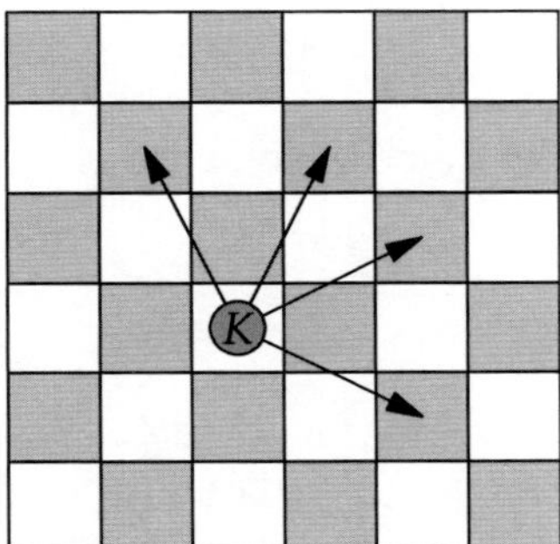

Figure 5.5

$\mathcal{W}_4$	$\mathcal{W}_4$	$\mathcal{L}_3$	$\mathcal{L}_3$	$\mathcal{W}_2$	$\mathcal{W}_2$	$\mathcal{L}_1$	$\mathcal{L}_1$
$\mathcal{W}_4$	$\mathcal{W}_4$	$\mathcal{L}_3$	$\mathcal{L}_3$	$\mathcal{W}_2$	$\mathcal{W}_2$	$\mathcal{L}_1$	$\mathcal{L}_1$
$\mathcal{W}_4$	$\mathcal{W}_4$	$\mathcal{W}_4$	$\mathcal{W}_4$	$\mathcal{W}_2$	$\mathcal{W}_2$	$\mathcal{W}_2$	$\mathcal{W}_2$
$\mathcal{W}_6$	$\mathcal{W}_4$	$\mathcal{W}_4$	$\mathcal{W}_4$	$\mathcal{W}_4$	$\mathcal{W}_2$	$\mathcal{W}_2$	$\mathcal{W}_2$
$\mathcal{W}_6$	$\mathcal{W}_6$	$\mathcal{L}_5$	$\mathcal{L}_5$	$\mathcal{W}_4$	$\mathcal{W}_4$	$\mathcal{L}_3$	$\mathcal{L}_3$
$\mathcal{W}_6$	$\mathcal{W}_6$	$\mathcal{L}_5$	$\mathcal{L}_5$	$\mathcal{W}_4$	$\mathcal{W}_4$	$\mathcal{L}_3$	$\mathcal{L}_3$
$\mathcal{W}_6$	$\mathcal{W}_6$	$\mathcal{W}_6$	$\mathcal{W}_6$	$\mathcal{W}_4$	$\mathcal{W}_4$	$\mathcal{W}_4$	$\mathcal{W}_4$
$\mathcal{L}_7$	$\mathcal{W}_6$	$\mathcal{W}_6$	$\mathcal{W}_6$	$\mathcal{W}_6$	$\mathcal{W}_4$	$\mathcal{W}_4$	$\mathcal{W}_4$

Figure 5.6

Now we need to determine whether $(0,0)$ is a $\mathcal{W}$ or an $\mathcal{L}$-position using the labelling rules. To simplify the diagram we will not draw the edges of the game graph. We will use the cells of the chessboard to represent the vertices. The final positions are precisely those with no followers, so we start by labelling these as $\mathcal{L}$-positions.

The convention that the player who cannot make a valid move loses is called the *normal play* convention, and we will use it from now on for games where no winning positions are explicitly defined.

The fully labelled chessboard is shown in figure 5.6, where, as before, the subscripts indicate the order in which the labels where added. We see that $(0,0)$ is an $\mathcal{L}$-position so Bob has a winning strategy.

Before discussing the next game we recall that a *polyomino* is a shape made up of unit squares called *cells* where each cell shares an entire edge with at least one other cell. A *domino* is a polyomino made up of two cells, and when we say that a domino is placed on another polyomino, we mean that the cells of the domino coincide exactly with two of the cells of the polyomino.

The game of *cram* is played on a polyomino (often a rectangular grid of cells), and the players take turns to place dominoes on the polyomino such that no two dominoes overlap. Under the normal play convention, the first player who is unable to place a domino loses.

Problem 5.5

Alice and Bob play cram on a 3×3 square grid. If Alice goes first, who has a winning strategy?

If we construct a game graph in a totally naïve way, it is impractically large. Fortunately we do not need the whole graph.

On Alice's first turn she has two options: she can cover one of the corner cells, or she can cover the centre cell.

If she starts by covering a corner, then Bob could respond by covering the centre and leaving five cells in a symmetrical L shape. This symmetry might appeal to Bob since it leaves Alice with only two different options to choose from. Moreover, if Alice plays in one 'arm' of the L, then Bob will be able to play in the other arm and win regardless of whether Alice covered the 'corner' of the L on her turn.

If, on the other hand, Alice starts by covering the centre cell, Bob can still create the five cell L shape and win as before.

At the end of this discussion we see that, if we want to use a game graph at all, we only need to consider the small section of the graph shown in figure 5.7.

This graph does not include all possible game positions. However, it does include every possible follower of the positions which are labelled $\mathcal{L}$. This is essential since the labelling rules only allow a position to be classified as $\mathcal{L}$ if all of its followers are $\mathcal{W}$.

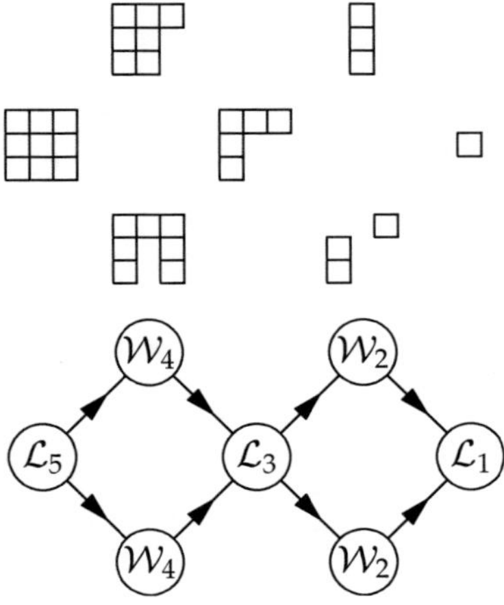

Figure 5.7

Exercise 5a

1. There is a pile of twenty-one counters on a table. Alice and Bob take it in turns to remove one, three or four counters from the pile. If Alice goes first, who has a winning strategy? (Since no winning condition is specified, we assume the normal play convention: the last player to move wins.)

2. There is a pile of twenty-one counters on a table. Alice and Bob take it in turns to remove one, three, five or seven counters from the pile. If Alice goes first, who has a winning strategy?

3. A counter is placed on the point $(7,7)$. Alice and Bob take it in turns to reduce one of the counter's coordinates by a whole number subject to the condition that neither coordinate can ever be negative. If Alice goes first, who has a winning strategy?

4. There is a pile of one hundred and ten counters on a table. Alice and Bob take it in turns to remove a (non-zero) square number of counters from the table. If Alice goes first, who has a winning strategy?

5.3 Symmetry and strategy stealing

There are a number of combinatorial games where, even if the full game graph is extremely complicated, one of the players can easily force a win by preserving some sort of symmetry in the game positions. We will study a few classic examples.

Problem 5.6

The game of *nim* is played with a number of piles of counters. Players take it in turns to choose a pile and take some counters from it. They must take at least one, and may take any number. They may even take the whole pile. Under the normal play convention the player who takes the last counter wins.

Alice and Bob play nim with a pile of n counters and a pile of m counters. If Alice goes first, for which values of m and n does she have a winning strategy?

We will write $[n, m]$ to represent the position with one pile of n and one pile of m counters.

By now we are used to the idea of working backwards from the final position of the game which is $[0, 0]$. We also remember to make the obvious observation that, if one of the piles contains no counters, then Alice can win on her first turn, so $[0, m]$ is a $\mathcal{W}$-position for all $m > 0$. The next simplest position is $[1, 1]$. This is a win for Bob, that is an $\mathcal{L}$-position, since Alice must move to $[0, 1]$ which is a $\mathcal{W}$-position.

We now see that $[1, m]$ is a $\mathcal{W}$-position for all $m > 1$, since Alice can always move to $[1, 1]$. This implies that $[2, 2]$ is an $\mathcal{L}$-position, which shows that $[2, m]$ is a $\mathcal{W}$-position for $m > 2$. At this point we might conjecture that $[n, m]$ is an $\mathcal{L}$-position if and only if $n = m$, and it is easy to see that this is indeed the case. We could express the argument using the labelling rules, but this is unnecessary. The player who starts facing the symmetrical position $[n, n]$ must move to an asymmetrical position $[n, m]$ where $n \neq m$. Their opponent can always restore symmetry, so is the only one who can move to the symmetrical position $[0, 0]$. Thus Alice has a winning strategy precisely when $n \neq m$. It is worth noting that the game in question 3 of exercise 5a is equivalent to playing nim with two piles of seven counters.

Problem 5.7

A position in the game of *Kayles* consists of a number of skittles arranged in a line, where the distances between skittles are all integers. Each turn the players (who are assumed to be expert bowlers) may knock over and remove any single skittle or pair of skittles that are one unit apart.

The initial position has n skittles placed at unit intervals, and Alice goes first. For which n does she have a winning strategy?

For $n = 1$ or 2 Alice can knock over all the skittles immediately and win. For $n = 3$ she wins by knocking over the middle skittle, and for $n = 4$ she can win by knocking over the middle two skittles. For $n = 5$ she also has a winning strategy. She starts by knocking over the middle skittle to leave two pairs of adjacent skittles. If Bob knocks down two skittles she can win immediately, and if he knocks down one, she knocks down one from the other pair. This leaves a pair of isolated skittles which is a win for Alice.

It is not too hard to see that Alice can win for any value of n. She simply knocks down the middle skittle for odd values of n and the middle pair of skittles for even values of n. This leaves Bob facing two identical rows of skittles. Whatever move he makes, Alice can mimic it in the other row, so Bob can never make the last move in the game.

This analysis did not involve the game graph at all, and was much quicker than using the labelling rules. The price we pay is that if the initial position cannot be made symmetrical in the first turn, then we have no idea who has a winning strategy. We will be able to remedy this shortcoming in section 5.6.

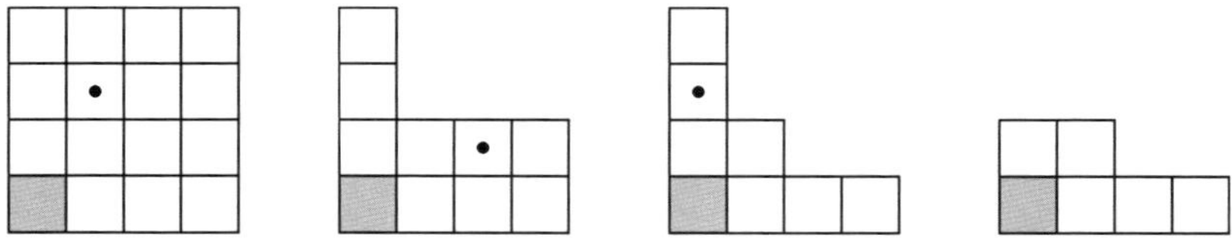

Figure 5.8

Problem 5.8

The game of *chomp* is played with an $n \times m$ rectangular chocolate bar consisting of square cells. We assume the centres of the cells have integer coordinates (x, y) where $1 \leq x \leq n$ and $1 \leq y \leq m$. The players take it in turns to take rectangular 'bites' out of the top right corner of the bar. That is, they choose a cell centred at (x_0, y_0) and remove all cells with centres (x, y) where $x_0 \leq x$ and $y_0 \leq y$. We will call this *chomping* (x_0, y_0).

Figure 5.8 shows a sequence of three moves on a 4×4 starting bar. Each turn the square containing the dot is 'chomped'.

The cell $(1, 1)$ is poisoned, so the player who takes it *loses* the game.

If Alice goes first, prove that she has a winning strategy unless $n = m = 1$.

This is a classic problem, and its solution is easy to follow, but hard to discover.

We use a proof by contradiction. Suppose that Bob has a winning strategy. That means that whatever move Alice makes first, Bob can respond by moving to an $\mathcal{L}$-position. In particular, if Alice chomps the single cell (n, m) then Bob can respond by moving to an $\mathcal{L}$-position. However, this position looks like the initial rectangle after a single chomp, so Alice could have moved to this $\mathcal{L}$-position herself.

This is called a *strategy stealing argument*. If Bob manages to find a good strategy, then Alice is able to steal it from him. The proof is striking because it does not give any indication of *how* Alice should go about winning; it merely proves that she can win if she plays perfectly. This is typical of proofs by contradiction, but it is still slightly frustrating. For a large starting board we know that the initial position is $\mathcal{W}$ and therefore has an $\mathcal{L}$-position among its followers, but neither we, nor anyone else, knows how to find such an $\mathcal{L}$-position.

Problem 5.9

Alice and Bob play a game on an $n \times m$ chessboard.

Initially all the cells are white. On Alice's first turn she paints a cell of her choice black. In subsequent turns the player chooses a white cell which is adjacent to exactly one black cell, and paints the chosen cell black. (Bob paints the second black cell, then Alice paints the third and so on.)

Under the normal play convention, determine who has a winning strategy in the following cases:

(a) $n = m$;

(b) n, m both odd;

(c) $m = 2$ and $n \geq 2$.

[Lomonosov Tournament 2013]

The conditions for the three parts give generous clues as to how to solve this problem. They also (accurately) suggest that this game may be hard to analyse for a general $n \times m$ chessboard.

For part (a) the solution is likely to hinge on some property enjoyed by square boards which other rectangular boards do not share. Symmetry in a diagonal is an obvious candidate, and it turns out that Alice can exploit this symmetry. She starts by choosing a diagonal D of the board and painting a cell which lies on D. She then uses her subsequent moves to ensure the board is always symmetric in D. Put another way, whichever cell Bob paints, Alice will paint the cell that is its reflection in D. This ensures Bob can never make the last move unless he is able to paint a cell on D. Fortunately for Alice, this is impossible: before Bob's turn the board is symmetric, so if a cell on D is adjacent to a black cell, then it must be adjacent to at least two black cells, and therefore cannot be painted.

For part (b) we are looking for a property which is peculiar to odd by odd boards. Here the key observation is that an odd by odd board has

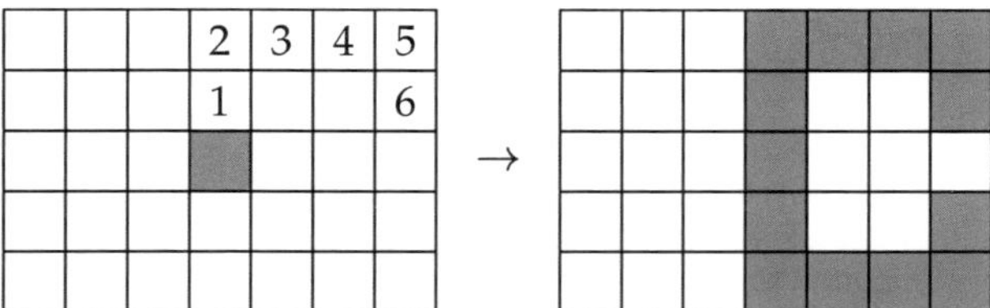

Figure 5.9

a unique central cell. Alice can use this to her advantage. She starts by painting the central cell, and then promises to use her subsequent turns to ensure the position is symmetrical in some way.

For example, Alice might try to preserve line symmetry in the central row. If Bob paints a cell which is not in the middle row, Alice can respond by painting its mirror image, while if Bob paints a cell in the middle row, Alice can preserve the symmetry by painting another cell in that row. Since the number of cells in the central row is odd, Alice might think that she will never be left without a move. However, this argument contains a subtle flaw, since Bob can use his turns to ensure that one of the cells in the central row is unavailable to Alice. If he manages to do this then he may be able to force a win. This problem is illustrated on a 7×5 grid in figure 5.9. If Bob uses his first six turns to paint the squares labelled 1 to 6 on the left, then if Alice preserves symmetry in the central row the grid will end up painted as shown on the right. Now there are only five more squares available to paint on the central row, so Bob paints these in turn Alice will eventually be left without a response.

A more successful strategy is for Alice to ensure that after each of her turns the board has rotational symmetry of order at least 2 about its central square. Alice starts by painting the central square. After this she always paints the square diametrically opposite the square just painted by Bob. This square can never share a border with the one just painted by Bob, so Alice will have a legal response to every move Bob makes and will therefore win the game in the end.

We rightly expect part (c) to be the hardest. There do not seem to be any obvious symmetries which are unique to $n \times 2$ boards, so we may need to work a little harder for a solution.

On an $n \times 2$ board there are essentially two different types of move: either paint the first black cell in a column, which we will call *opening* that column, or paint the second black cell in a column, which we will call *filling* that column.

As always, it is worth studying small values of n.

If $n = 2$ the game is sure to last three turns so Alice wins.

If $n = 3$ the game will either last four or five turns depending on whether or not the central column gets filled. It is in Alice's interest to prevent this from happening, since then the game will last five turns. She can start by opening the first column from the left. If Bob opens the second column, she can fill the first and vice versa. Either way she is sure to win.

If $n = 4$ it is worth thinking about the possible finishing positions for the game. Certainly every column will contain at least one black cell, and, if fewer than two columns have been filled, then there is sure to be another available move. Moreover, it is impossible to have three filled columns so the game will last exactly six turns. In other words, Bob is sure to win no matter how he plays.

It is becoming clear that the key thing is the number of filled columns in the finishing position for the game. We will call this number F. If F has the same parity as n, then the total number of black cells is even so Bob has won the game.

If $n = 5$ then Alice's only hope is to ensure that $F = 2$. To make this happen it is enough to ensure that the second column is filled, and Alice can guarantee this will happen within the first three turns of the game.

For $n = 6$ Alice would like F to equal three. The only way this can fail to happen is if the second and fifth columns are filled. Since Alice can ensure that the first column is filled within the first three turns, she can force a win.

For $n = 7$ Alice would like F to equal four. She can certainly start by ensuring the first column is filled, but she also needs to ensure that the finishing position does not contain two adjacent unfilled columns. Fortunately wide gaps like this are not too hard to prevent. Alice simply fills up columns whenever she can. Bob can also prevent there from being adjacent unfilled columns in the final position if he so wishes.

This allows us to give a complete solution to the problem.

If n is odd Alice has a winning strategy. She promises that the final position will not have adjacent unfilled columns. This means that either every even-numbered column or every odd-numbered column will be filled in the final position. Since Alice can open a column of her choice and ensure it is filled within the first three turns, she can control the parity of F. In particular she can ensure it is even.

If $n = 4k + 2$ Alice also has a winning strategy. By promising that the final position will not contain adjacent unfilled columns, she can ensure that $F = 2k + 1$.

If $n = 4k$ Bob has a winning strategy. He promises that the final position will not contain adjacent unfilled columns, which ensures that $F = 2k$.

Exercise 5b

1. The number 10^{2007} is written on a blackboard. Alice and Bob take it in turns to perform one of the following operations:

(a) replace a number x on the board with two integers $a, b > 1$ such that $ab = x$;

(b) choose two equal numbers on the board and erase either one or both of them.

If Alice goes first, who has a winning strategy under the normal play convention? [Nordic Mathematical Contest 2007]

2. Alice and Bob play cram on a 3×4 grid. Who has a winning strategy? Can you generalise to a $(2n + 1) \times 2m$ grid?

3. Alice and Bob play cram on a 4×4 grid. Who has a winning strategy? Can you generalise to a $2n \times 2m$ grid?

4. Alice and Bob take it in turns writing factors of 160 000 on a board. It is illegal to write a number which is a multiple of another number which is already on the board, and the player who writes '1' loses. If Alice goes first, who has a winning strategy?

5. Alice and Bob want to play chomp on a chocolate bar which is infinite in both directions, that is one whose cells are centred on (x, y) for any $x, y > 0$.

(a) Is it possible for this game to go on forever?

(b) Which player(s) can be certain not to lose?

6. Let n be a positive integer. Two players A and B play a game in which they take turns choosing positive integers $k \leq n$.
The rules of the game are:
   (i) a player cannot choose a number that has been chosen by either player on any previous turn.
   (ii) a player cannot choose a number consecutive to any of those the player has already chosen on any previous turn.
   (iii) the game is a draw if all numbers have been chosen; otherwise the player who cannot choose a number anymore loses the game.

   The player A takes the first turn. Determine the outcome of the game, assuming that both players use optimal strategies.
   (Note: this is actually a partisan game.)
   [International Mathematical Olympiad shortlist 2015]

5.4 Contest problems

We will now spend some time discussing a particularly challenging problem from the 2015 Romanian Masters of Mathematics contest.

Problem 5.10

For an integer $n \geq 5$, two players play the following game on a regular n-gon. Initially, three consecutive vertices are chosen, and one counter is placed on each. A move consists of sliding one counter along any number of edges to another vertex of the n-gon without jumping over another counter. A move is *legal* if the area of the triangle formed by the counters is strictly greater after the move than before. The players take turns to make legal moves, and if a player cannot make a legal move, that player loses.

For which values of n does the player making the first move have a winning strategy? [Romanian Masters of Mathematics 2015]

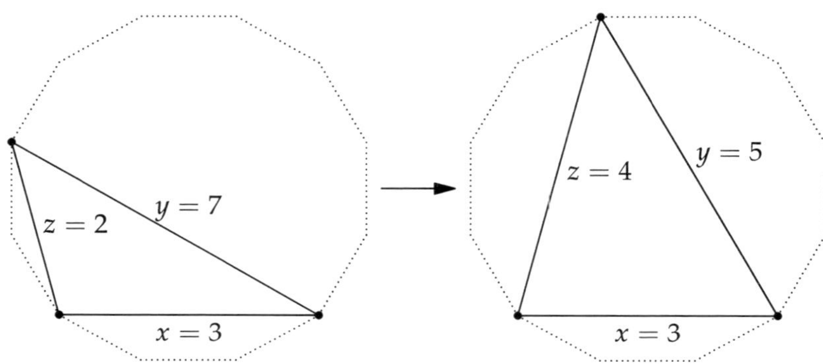

Figure 5.10

The first order of business is to find good notation. We might represent a position by the triple (x, y, z) where x, y and z are the number of edges between pairs of counters going, say, anticlockwise round the n-gon. Clearly triples obtained by permuting the coordinates, like (x, z, y) and (y, z, x) represent congruent triangles and are therefore *equivalent*. Thus, for example, the starting position for this game is equivalent to $(1, 1, n-2)$.

There is some redundancy in this notation since we always have that $x + y + z = n$, so x and y determine z. However, listing all three numbers makes the moves in the game easier to represent. In particular, any move, legal or otherwise, consists of choosing two of the members of the set and replacing them with two other positive integers with the same sum.

How are we to distinguish legal moves from illegal ones in this representation? Our first thought might be to try and express the area of the triangle formed by the counters in terms of x, y and z. This is possible, but to call the calculation unappealing would be something of an understatement. Fortunately, the exact area of the triangle formed by the counters does not matter, only whether or not a move increases it. Figure 5.10 shows the move $(3, 7, 2) \to (3, 5, 4)$ played on a dodecagon.

The diagram makes it clear that this is, in fact, a legal move. The two triangles share a common base, labelled $x = 3$, but the move increases the height of the triangle since 5 and 4 are more nearly equal than 7 and 2. More generally we see that a legal move consists of choosing two numbers in the triple and replacing them with two integers with the same sum and a strictly smaller difference. To put it another way, the legal moves in this game make the numbers x, y and z more nearly equal.

With this in place we can begin analysing the game for some small values of n to see if we can classify the $\mathcal{L}$ and $\mathcal{P}$-positions.

When $n = 3$, $(1,1,1)$ is the only position so this is $\mathcal{L}$, and when $n = 4$, $(1,1,2)$ is also $\mathcal{L}$.

When $n = 5$ we have that $(1,2,2)$ is the final position so this is $\mathcal{L}$. We can reach this position immediately from $(1,1,3)$ so $(1,1,3)$ is a $\mathcal{W}$-position.

When $n = 6$ the final ($\mathcal{L}$) position is $(2,2,2)$. This can be reached from $(1,2,3)$ so $(1,2,3)$ is $\mathcal{W}$. The initial position $(1,1,4)$ can only be followed by $(1,2,3)$, so $(1,1,4)$ is another $\mathcal{L}$ position.

When $n = 7$ we see that $(2,2,3)$ is $\mathcal{L}$ which implies that $(1,2,4)$ and $(1,3,3)$ are $\mathcal{W}$. These are the only followers of $(1,1,5)$ so $(1,1,5)$ is an $\mathcal{L}$-position.

When $n = 8$ we have that $(2,3,3)$ is $\mathcal{L}$ so $(1,2,5)$, $(1,3,4)$ and $(2,2,4)$ are all $\mathcal{P}$. The initial position $(1,1,6)$ is followed by the first two of these $\mathcal{P}$-positions, so it is $\mathcal{L}$.

So far we have found the following $\mathcal{L}$-positions:

$$(1,1,1),(1,1,2),(1,2,2),(2,2,2),(1,1,4),(2,2,3),(1,1,5),(2,3,3),(1,1,6).$$

We notice that all these positions are of the form (a,a,b) where a may or may not be less than b. Such positions represent isosceles triangles on the n-gon, so we will call them *isosceles positions*.

It certainly is not the case that the $\mathcal{L}$-positions are precisely the isosceles positions since $(1,1,3)$, $(1,3,3)$ and $(2,2,4)$ are all $\mathcal{W}$, but it may well be that the $\mathcal{L}$-positions are a subset of the isosceles positions. Testing a few more small values of n provides more support for this conjecture, so it is worth temporarily assuming that it holds and exploring its consequences.

If every $\mathcal{L}$-positions is isosceles, then the position (a,a,b) can only fail to be $\mathcal{L}$ if one of its followers is an isosceles position which is $\mathcal{L}$. We therefore need to understand when (a,a,b) has isosceles followers. If $a = b$ then the position has no followers at all so must be $\mathcal{L}$, and if $a \neq b$ then the only possible isosceles follower of (a,a,b) is $(a, \frac{a+b}{2}, \frac{a+b}{2})$. This position is reached by moving the counter on the original n-gon along $|a - \frac{a+b}{2}| = |\frac{1}{2}(a-b)|$ edges.

Thus, if we have a the position (a,a,b), then we focus on the quantity $d = |a-b|$. If d is zero or an odd number, then the position has no isosceles followers so must be $\mathcal{L}$.

If $a-b$ is even, then (a,a,b) is followed by $(a, \frac{a+b}{2}, \frac{a+b}{2})$. For this position $d = \frac{1}{2}|\frac{a+b}{2} - a| = \frac{1}{4}|a-b|$ which is exactly half what it was before.

The consequence of this is that if $|a-b|$ is even but not divisible by four then (a, a, b) is $\mathcal{P}$ since it is followed by an $\mathcal{L}$-position.

More generally if the difference between the numbers in an isosceles position is even (and non-zero), then we can move to exactly one other isosceles position and halve the relevant difference in the process. We can keep doing this until the difference reaches an odd number at which point we have an isosceles position with no isosceles followers. This means that in order to predict whether a given isosceles position (a, a, b) is an $\mathcal{L}$ or $\mathcal{P}$-position we need to know how many times $a-b$ can be halved before reaching an odd number (unless $a = b$ in which case the position is $\mathcal{L}$).

Let us define $v_2(n)$ to be the exponent of 2 in the prime factorisation of an integer n. For example, $v_2(46) = 1$, $v_2(47) = 0$ and $v_2(48) = 4$.

We are now in a position to formulate our conjecture as to the nature of the positions exactly.

Claim

A position is $\mathcal{L}$ precisely if it is of the form (a, a, b) where either $a = b$ or $v_2(|a-b|)$ is even.

We have done most of the work already, but our arguments so far rest on the assumption that every $\mathcal{L}$-position is isosceles. To prove the claim, we must check that our proposed labelling of the positions satisfies the position labelling rules.

If $n = 3k$ then the final position is (k, k, k) which we label $\mathcal{L}$. If $n = 3k+1$ or $3k+2$ the final positions are $(k, k, k+1)$ and $(k, k+1, k+1)$ respectively. To determine the labels of these positions we must consider $v_2(|k+1-k|)$ which is zero. Since zero is even, we see that final positions are always labelled $\mathcal{L}$, just as they should be.

It is easy to see that none of the claimed $\mathcal{L}$-positions follow each other. If two isosceles positions follow each other then they are of the form (a, a, b) and $(a, \frac{a+b}{2}, \frac{a+b}{2})$. Exactly one of $v_2(|a-b|)$ and $v_2(|\frac{a-b}{2}|)$ is even, so exactly one of these positions is $\mathcal{L}$.

Finally we must check that every position we claim is $\mathcal{P}$ has an $\mathcal{L}$-position among its followers. The previous remark makes this clear for isosceles $\mathcal{P}$-positions, so it remains to check that every non-isosceles position is followed by an $\mathcal{L}$-position.

If we consider the position (x, y, z) where we assume $x < y < z$, then we can replace the x and z with y and $(x-y+z)$ to reach the isosceles position $(y, y, x-y+z)$. To determine the label of this position we consider $v_2(|x-2y+z|)$. If this is even we are done, so we are left with the case where $v_2(|x-2y+z|)$ is odd. In this case $x-2y+z$ has at least one two among its prime factors, which implies that $x+z$ is even. However, if $x+y$ is even, then we may start with the position (x,y,z) and replace x and z with two copies $\frac{x+z}{2}$ to reach the isosceles position $(y, \frac{x+z}{2}, \frac{x+z}{2})$. To determine the label of this position we consider $v_2(|y-\frac{x+Z}{2}|) = v_2(\frac{1}{2}|x-2y+z|)$. Now the fact that $v_2(|x-2y+z|)$ is odd implies that $(y, \frac{x+z}{2}, \frac{x+z}{2})$ is an $\mathcal{L}$-position and the proof is complete.

This problem comes from perhaps the hardest international high school maths competition, so it is not surprising that the solution was tough. However, with hindsight it should not be surprising that isosceles positions are of crucial interest. These positions are a natural thing to focus our attention on since fewer moves are available from isosceles positions than completely general ones. We can turn this into a slogan for problems about games:

Focus on positions which limit the options for the next player.

Exercise 5c

1. A chocolate bar is made of n^2 equilateral triangles of side length 1, arranged to form an equilateral triangle of side length $n \geq 2$. Alice and Bob take it in turns breaking off pieces of chocolate from the corners of the bar and eating them. Each piece must be an equilateral triangle with integer side length, and neither player may finish the bar. (Two possible games for $n = 3$ are shown in figure 5.11.)

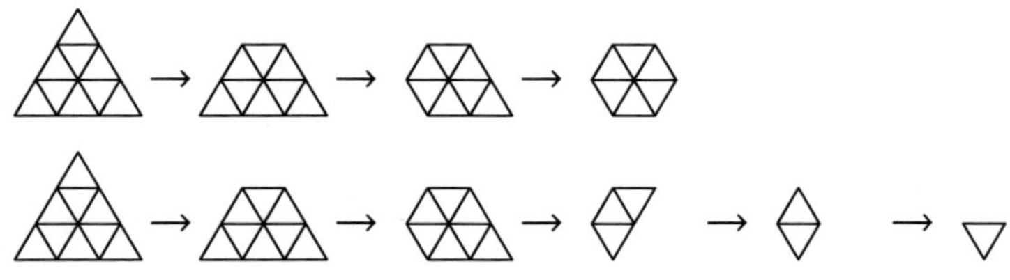

Figure 5.11

For each n, find who has a winning strategy under the normal play convention if Alice goes first. [Tournament of the Towns 2003]

2. There are $n > 2$ piles each consisting of a single nut. Alice and Bob take it in turns to combine two piles which contain coprime numbers of nuts into a new pile. If Alice goes first, who has a winning strategy under the normal play convention? (Recall that two numbers are coprime if if their highest common factor is 1.)
   [Tournament of the Towns 2008]

3. There is a pile of s stones on a table. Alice and Bob take it in turns to remove stones from the pile. On each turn they may take one stone, a prime number of stones, or a positive multiple of n stones, where n is a fixed positive integer. The game is played with the normal play convention, and Alice goes first.
   For each n, find the number of values of s for which Bob has a winning strategy. [Junior Balkan Mathematical Olympiad 2014]

4. Naomi and Tom play a game, with Naomi going first. They take it in turns to pick an integer from 1 to 100, each time selecting an integer which no-one has chosen before. A player loses the game if, after their turn, the sum of all the integers chosen since the start of the game (by both of them) cannot be written as the difference of two square numbers. Determine if one of the players has a winning strategy, and if so, which. [British Mathematical Olympiad Part One 2016]

5. Alice and Bob play the following game. They start with two non-empty piles of coins. Taking turns, with Alice playing first, each player chooses a pile with an even number of coins and moves half of the coins of this pile to the other pile. The game ends if a player cannot move, in which case the other player wins. Determine all pairs (a, b) of positive integers such that if initially the two piles have a and b coins, respectively, then Bob has a winning strategy.
   [Balkan Mathematical Olympiad 2018]

5.5 Nim

We start this section by recalling that a position in the game of *nim* consists of a number of piles of counters, and that a turn consists of removing a positive number of counters from any one pile. With only two piles of counters the game is very straightforward: the $\mathcal{L}$-positions are precisely those where the piles are equal in size. When there are more than two piles, the game is rather more interesting, and it is worth spending some time learning how to play it perfectly.

Problem 5.11

Nim is often played with an initial position consisting of a pile of three, a pile of five and a pile of seven counters. Is this position a $\mathcal{W}$ or an $\mathcal{L}$-position?

Rather than attack this problem directly, we will try and build up a general understanding of the $\mathcal{L}$-positions in nim with three piles. We will write $[x, y, z]$ for the position where the piles contain x, y and z counters respectively. The order of the numbers does not matter, and we will often (though not always) choose to have $x \leq y \leq z$.

If the smallest pile contains no counters we know that $[0, y, y]$ is an $\mathcal{L}$-position for all y, and that $[0, y, z]$ is a $\mathcal{W}$-position if $y \neq z$.

Now we look at positions where the smallest pile has one counter, which we will order by the size of the second smallest pile. The position $[1, 1, z]$ is a $\mathcal{W}$-position for $z \geq 1$ since the next player can move to $[1, 1, 0]$. We will denote this by writing $[1, 1, z] \to [1, 1, 0]$.

The fact that $[1, 2, 2] \to [0, 2, 2]$ shows that $[1, 2, 2]$ is a $\mathcal{W}$-position.

Now we come to the position $[1, 2, 3]$. We claim that this is an $\mathcal{L}$-position, that is, that all of its followers are $\mathcal{W}$-positions.

To check this we list the six followers of $[1, 2, 3]$, together with the $\mathcal{L}$-positions which can be reached from each of them.

$$[0,2,3] \to [0,2,2] \qquad [1,2,2] \to [0,2,2]$$
$$[1,1,3] \to [1,1,0] \qquad [1,2,1] \to [1,0,1]$$
$$[1,0,3] \to [1,0,1] \qquad [1,2,0] \to [1,1,0]$$

This means that if $z > 3$ then $[1, 2, z]$ is a $\mathcal{W}$-position since the move $[1, 2, z] \to [1, 2, 3]$ can be made.

We also have that $[1,3,z]$ is a $\mathcal{W}$-position (provided $z > 2$) since $[1,3,z] \to [1,3,2]$ and $[1,3,2]$ is equivalent $[1,2,3]$.

Next we consider positions of the form $[1,4,z]$ for $z \geq 4$. We have that $[1,4,4] \to [0,4,4]$ so $[1,4,4]$ is $\mathcal{W}$. We claim that $[1,4,5]$ is an $\mathcal{L}$-position. Just as we did with the position $[1,2,3]$ we check that an $\mathcal{L}$-position can be reached from every follower of $[1,4,5]$. This check is shown below.

$$\begin{array}{ll} [0,4,5] \to [0,4,4] & \quad [1,4,4] \to [0,4,4] \\ [1,3,5] \to [1,3,2] & \quad [1,4,3] \to [1,2,3] \\ [1,2,5] \to [1,2,3] & \quad [1,4,2] \to [1,3,2] \\ [1,1,5] \to [1,1,0] & \quad [1,4,1] \to [1,0,1] \\ [1,0,5] \to [1,0,1] & \quad [1,4,0] \to [1,1,0] \end{array}$$

If $z > 5$ then $[1,4,z] \to [1,4,5]$, and if $z > 4$ then $[1,5,z] \to [1,5,4]$, so $[1,4,z]$ and $[1,5,z]$ are all $\mathcal{W}$-positions.

Arguing in a similar way, we see that $[1,6,6]$ is a $\mathcal{W}$-position and that $[1,6,7]$ is an $\mathcal{L}$-position. Indeed, it is easy to check that a nim position with a pile containing one counter is an $\mathcal{L}$-position if, and only if, it is of the form $[1,E,E+1]$ where E is any even number. You should take a moment to convince yourself that no follower of $[1,E,E+1]$ has the form $[1,E',E'+1]$ for some other even number E', and also that if $[1,y,z]$ is not of the form $[1,E,E+1]$ then one of its followers is (assuming $y \neq z$).

Now we are ready to tackle positions where the smallest pile contains two counters.

$$\begin{array}{l} [2,2,z] \to [2,2,0] \text{ for } z \geq 2 \\ [2,3,z] \to [2,3,1] \text{ for } z \geq 3 \\ [2,4,4] \to [0,4,4] \\ [2,4,5] \to [1,4,5] \end{array}$$

So the positions on the left are all $\mathcal{W}$-positions.

The next position to consider is $[2,4,6]$. The list below shows that every follower of this position is a $\mathcal{W}$-position. So $[2,4,6]$ is an $\mathcal{L}$-position.

$[1,4,6] \to [1,4,5]$	$[2,4,5] \to [1,4,5]$
$[0,4,6] \to [0,4,4]$	$[2,4,4] \to [0,4,4]$
$[2,3,6] \to [2,3,1]$	$[2,4,3] \to [2,1,3]$
$[2,2,6] \to [2,2,0]$	$[2,4,2] \to [2,0,2]$
$[2,1,6] \to [2,1,3]$	$[2,4,1] \to [2,3,1]$
$[2,0,6] \to [2,0,2]$	$[2,4,0] \to [2,2,0]$

We have $[2,4,z] \to [2,4,6]$ for $z > 6$ and $[2,6,z] \to [2,6,4]$ for $z > 4$ so these are all $\mathcal{W}$-positions.

We also have $[2,5,5] \to [0,5,5]$ and $[2,5,6] \to [2,4,6]$, so these are also $\mathcal{W}$. However, you should take a moment to check systematically that $[2,5,7]$ is an $\mathcal{L}$-position since all its followers are $\mathcal{W}$.

At this point we are able to present a solution to the original problem. We see that $[3,5,7]$ is a $\mathcal{W}$-position. If Alice goes first she can win by moving to $[2,5,7]$ on her first turn.

This is good news, but it is worth exploring the $\mathcal{L}$-positions a little further.

Problem 5.12

Find necessary and sufficient conditions for $[x,y,z]$ to be an $\mathcal{L}$-position in nim.

Our first instinct might be to look at all the $\mathcal{L}$-positions we have found so far and try to spot some patterns. This is a good idea, but we should think carefully about how we represent these positions. One option is simply to write a list, but it is not immediately obvious which order we ought to list the positions in. An alternative is to allow the triple $[x,y,z]$ to remind us of the triple (x,y,z). This in turn reminds us of coordinates which suggests slightly new ways of thinking about the problem.

We can view the set of $\mathcal{L}$-positions as a particular set of points in three-dimensional space. However, since we are working on paper, we need a two-dimensional representation of these points, so we try the following. At each point with coordinates (x,y) we write the value of z which makes $[x,y,z]$ an $\mathcal{L}$-position. This provides a sort of map of the points in three

9	9	8	?	?	?	?	?	?	1	0
8	8	9	?	?	?	?	?	?	0	1
7	7	6	5	?	?	2	1	0	?	?
6	6	7	4	?	2	?	0	1	?	?
5	5	4	7	?	1	0	?	2	?	?
4	4	5	6	?	0	1	2	?	?	?
3	3	2	1	0	?	?	?	?	?	?
2	2	3	0	1	6	7	4	5	?	?
1	1	0	3	2	5	4	7	6	9	8
0	0	1	2	3	4	5	6	7	8	9
	0	1	2	3	4	5	6	7	8	9

Table 5.1

dimensions. We do not yet know whether every pair (x, y) has such a z-value associated with it, so our map may contain gaps. However, we do know that there can be at most one z-value for each pair (x, y) since as soon as we find a z_0 such that $[x, y, z_0]$ is $\mathcal{L}$ we know that $[x, y, z]$ is $\mathcal{W}$ for all $z > z_0$.

The $\mathcal{L}$-positions we have found so far are shown in table 5.1.

There are still too many unknown values to see many patterns, so our next task is to find out which value of z (if any) makes $[3, 4, z]$ an $\mathcal{L}$-position. The following observations show that $z = 7$ is the first plausible candidate and it is easy to check that this is indeed an $\mathcal{L}$-position.

$$[3,4,0] \to [3,3,0] \qquad [3,4,4] \to [0,4,4]$$
$$[3,4,1] \to [3,2,1] \qquad [3,4,5] \to [1,4,5]$$
$$[3,4,2] \to [3,1,2] \qquad [3,4,6] \to [2,4,6]$$
$$[3,4,3] \to [3,0,3]$$

At this point we might conjecture that every $\mathcal{L}$-position is of the form $[x, y, x + y]$ for some x and y. We have a fair amount of evidence for this, so let us see how it stands up to the next test in our table. Which z-value makes $[3, 5, z]$ an $\mathcal{L}$-position? We start by noting that z is at least 6.

$$[3,5,0] \to [3,3,0] \qquad [3,5,3] \to [3,0,3]$$
$$[3,5,1] \to [3,2,1] \qquad [3,5,4] \to [1,5,4]$$
$$[3,5,2] \to [3,1,2] \qquad [3,5,5] \to [0,5,5]$$

9	9	8	?	?	?	?	?	?	1	0
8	8	9	?	?	?	?	?	?	0	1
7	7	6	5	4	3	2	1	0	?	?
6	6	7	4	5	2	3	0	1	?	?
5	5	4	7	6	1	0	3	2	?	?
4	4	5	6	7	0	1	2	3	?	?
3	3	2	1	0	7	6	5	4	?	?
2	2	3	0	1	6	7	4	5	?	?
1	1	0	3	2	5	4	7	6	9	8
0	0	1	2	3	4	5	6	7	8	9
	0	1	2	3	4	5	6	7	8	9

Table 5.2

It is straightforward to check that every follower of $[3,5,6]$ is a $\mathcal{W}$-position so $[3,5,6]$ is an $\mathcal{L}$-position. Our conjecture lies in tatters before us, and we begin to sense we may have underestimated this harmless sounding game.

Our two new $\mathcal{L}$-positions allow us to fill in a dozen new values in our table and we obtain table 5.2. Perhaps the most striking thing now is the long diagonal of 7s. This is interesting, but it is still far from obvious what is going on.

Another important observation is that our partially completed table appears to be a *Latin square*: no number occurs twice in the same row or column. Fortunately, this second observation is easier to understand. If $[x,y,z]$ is $\mathcal{L}$, then clearly $[x,y',z]$ cannot be $\mathcal{L}$ for any $y' > y$, so the same z-value can never occur at two different points in the same column, and similarly for the rows. This is real progress because it makes populating other cells in the table much quicker.

For example, suppose we want find z such that $[2,8,z]$ is $\mathcal{L}$. Reading up the $x = 2$ column we see that z cannot be 2, 3, 0, 1, 6, 7, 4 or 5, and reading across the row we see that it cannot be 8 or 9. Thus the first plausible candidate is $z = 10$.

Might $[2,8,10]$ have an $\mathcal{L}$-position among its followers? Clearly taking counters from the first pile cannot lead to an $\mathcal{L}$-position because moving to $[x,8,10], x < 2$ corresponds to moving to the left in the table, and the number 10 does not occur in the $y = 8$ row for $x < 2$. Similarly taking counters from the pile of eight counters corresponds to moving down the

$x = 2$ column and $[2, y, 10]$ is not $\mathcal{L}$ for $y < 8$ since 10 does not occur below $y = 8$ in that column. The only other option is to take from the pile of ten counters, moving to $[2, 8, z]$. However, we have already checked that $[2, 8, z]$ is not $\mathcal{L}$ for $z < 10$, that was precisely why we focused on $z = 10$ in the first place.

The preceding two paragraphs can be easily adapted to give a general method for populating the table. Suppose we have a vacant cell (x_0, y_0) and that we are trying to find z_0 such that $[x_0, y_0, z_0]$ is $\mathcal{L}$. Suppose further that we have filled the cells directly below (x_0, y_0) and those directly to the left of (x_0, y_0). The z-values in the filled cells form a set $Z = \{z_1, z_2, \ldots z_n\}$. We claim that if we let z_0 be the least non-negative integer *not* in Z, then $[x_0, y_0, z_0]$ is $\mathcal{L}$.

We need to check that $[x_0, y_0, z_0]$ has no $\mathcal{L}$-positions among its followers. Taking counters from the first pile corresponds to moving to the cell (x, y_0) in the table (where $x < x_0$) and the z-value which makes this an $\mathcal{L}$-position is a member of Z so does not equal z_0. The same argument applies to taking counters from the second pile. Finally, if we take counters from the third pile we reach $[x_0, y_0, z]$ where $z < z_0$. Since z_0 is the least number not in Z we have that $z \in Z$. This shows that $[x_0, y_0, z]$ cannot be an $\mathcal{L}$-position since the same value cannot occur twice in a row or column of the table.

Put more succinctly, if Z is the set of values found below or left of the cell (x_0, y_0) then (x_0, y_0) contains the number $z_0 = mex(Z)$ where the *mex* of a set is the *minimum excluded element*.

With this in place it is fairly easy to fill in a large portion of the table. Table 5.3 shows the values for $x, y \leq 15$. It reveals a wealth of intriguing structure, and you should take a moment to study it.

We see long diagonal runs of 7s, 8s and 15s, and long increasing and decreasing sequences in the seventh, eighth and fifteenth rows. These patterns repeat on a smaller scale with 3s and 4s and also, to some extent, with 11s and 12s. Understanding everything at a glance is probably still asking too much, but one thing should certainly stand out: there is something going on with powers of 2.

In order to make further progress we need to think again about our notation. If powers of two are important, then we would like notation which makes powers of two particularly easy to work with. This suggests that we might try rewriting some of the table using *binary*. We will use the following notation: since $11 = 1 \times 8 + 0 \times 4 + 1 \times 2 + 1 \times 1$ we will write $11 =_2 1011$ and refer to the columns of this number as the '8s' column, the '4s' column and so on.

15	15	14	13	12	11	10	9	8	7	6	5	4	3	2	1	0
14	14	15	12	13	10	11	8	9	6	7	4	5	2	3	0	1
13	13	12	15	14	9	8	11	10	5	4	7	6	1	0	3	2
12	12	13	14	15	8	9	10	11	4	5	6	7	0	1	2	3
11	11	10	9	8	15	14	13	12	3	2	1	0	7	6	5	4
10	10	11	8	9	14	15	12	13	2	3	0	1	6	7	4	5
9	9	8	11	10	13	12	15	14	1	0	3	2	5	4	7	6
8	8	9	10	11	12	13	14	15	0	1	2	3	4	5	6	7
7	7	6	5	4	3	2	1	0	15	14	13	12	11	10	9	8
6	6	7	4	5	2	3	0	1	14	15	12	13	10	11	8	9
5	5	4	7	6	1	0	3	2	13	12	15	14	9	8	11	10
4	4	5	6	7	0	1	2	3	12	13	14	15	8	9	10	11
3	3	2	1	0	7	6	5	4	11	10	9	8	15	14	13	12
2	2	3	0	1	6	7	4	5	10	11	8	9	14	15	12	13
1	1	0	3	2	5	4	7	6	9	8	11	10	13	12	15	14
0	0	1	2	3	4	5	6	7	8	9	10	11	12	13	14	15
	0	1	2	3	4	5	6	7	8	9	10	11	12	13	14	15

Table 5.3

Table 5.4 shows the binary values for $x, y \leq 8$.

More patterns start to emerge. It seems that if we read up a column, the final digits alternate between 0 and 1, the penultimate digits come in alternating blocks of length two and the digits before that come in alternating blocks of length four. Let us see whether these patterns (which we have not yet proved persist throughout the table) allow us to predict z-values given the values of x and y.

First we focus on the final binary digits of x, y and z. Our table suggests that if the final digits of x and y are the same then z will end in a 0, while if they are different it will end in a 1.

Now we look at the second last digits. The table suggests exactly the same rule for these: z has a zero for its second last digit precisely when x and y have the same second last digit. Indeed, this rule seems to hold for the third last binary digit as well, so we conjecture that it holds for every binary digit.

Before proving our conjecture we will rephrase it one more time. In its current form there is one privileged pile called z whose size is determined by x and y. It would be more elegant to have a formulation where the

1000	1000	1001	1010	1011	1100	1101	1110	1111	0000
0111	0111	0110	0101	0100	0011	0010	0001	0000	1111
0110	0110	0111	0100	0101	0010	0011	0000	0001	1110
0101	0101	0100	0111	0110	0001	0000	0011	0010	1101
0100	0100	0101	0110	0111	0000	0001	0010	0011	1100
0011	0011	0010	0001	0000	0111	0110	0101	0100	1011
0010	0010	0011	0000	0001	0110	0111	0100	0101	1010
0001	0001	0000	0011	0010	0101	0100	0111	0110	1001
0000	0000	0001	0010	0011	0100	0101	0110	0111	1000
	0000	0001	0010	0011	0100	0101	0110	0111	1000

Table 5.4

symmetry between the three piles was more obvious. If we consider, say, the last binary digits of x, y and z, then our conjecture is that there will be either two 1s or no 1s at all. The same applies to the other binary digits, giving us the following symmetric statement of our conjecture:

Nim conjecture

> The position $[x, y, z]$ is an $\mathcal{L}$-position if, and only if, when we write the numbers x, y, z in binary one above the other, we find an even number of 1s in each column.

For example, suppose that $[x, y, z] = [5, 24, 29]$.

We have $5 = 4 + 1,\ 24 = 16 + 8$ and $29 = 16 + 8 + 4 + 1$, so writing these numbers above each other in binary gives:

$$5 =_2 00101$$
$$24 =_2 11000$$
$$29 =_2 11101$$

We find an even number of 1s in each column, so expect this to be an $\mathcal{L}$-position.

First we check that no followers of a conjectured $\mathcal{L}$-position satisfy our $\mathcal{L}$ criterion. This turns out to be easy to see. If we reduce a number, we change some of its binary digits. However, changing any binary digit changes the parity of the number of ones in its column. Therefore, if every column contained an even number of 1s before the change, then at least one column contains an odd number of 1s after the change.

Next we check that every position which does not satisfy our supposed $\mathcal{L}$-position criterion is followed by a position that does. This is easiest to see with an example. Suppose Alice makes the move $[5,24,29] \to [5,23,29]$. We suspect that Bob can respond by moving to an $\mathcal{L}$-position.

We represent the position $[5,23,29]$ in binary, and describe Bob's move in stages.

$$\begin{aligned} 5 &=_2 00101 \\ 23 &=_2 10111 \\ 29 &=_2 11101 \end{aligned}$$

- Working from the left, he finds the first column which contains an odd number of 1s.
  (In our example this is the second column from the left: the '8s' column.)
- He chooses a row which contains a 1 in that column.
  (In our example he must choose the bottom row.)
- He works from left to right along this row adding or removing counters so as to change the number of 1s in any column which originally contained an odd number of them.
  (In our example he removes eight counters, then he removes four more counters, then he returns two counters to the pile, and finally he removes one more counter. The net effect of this is that he removes $8+4-2+1=11$ counters.)

So Bob makes the move $[5,23,29] \to [5,23,18]$.
Writing $[5,23,18]$ in binary shows an even number of 1s in each column.

$$\begin{aligned} 5 &=_2 00101 \\ 23 &=_2 10111 \\ 18 &=_2 10010 \end{aligned}$$

As a second example, suppose Bob is faced with the position $[80,91,20]$.

$$\begin{aligned} 80 &=_2 1010000 \\ 91 &=_2 1011011 \\ 20 &=_2 0010100 \end{aligned}$$

The third column (the '16s') is the first containing an odd number of 1s. Bob can work with any of the rows in this representation, since all three contain a 1 in that column. If he works with the top row, he must remove sixteen, then add eight, then add four, then add two and then add one counter. Overall this amounts to the move $[80, 91, 20] \to [79, 91, 20]$.

$$\begin{aligned} 80 &=_2 1010000 & \qquad 79 &=_2 1001111 \\ 91 &=_2 1011011 & \qquad 91 &=_2 1011011 \\ 20 &=_2 0010100 & \qquad 20 &=_2 0010100 \end{aligned}$$

If Bob chooses to work with the second row, he must remove sixteen, then remove eight, then add four, then remove two and then remove one counter. Overall this amounts to $[80, 91, 20] \to [80, 68, 20]$.

$$\begin{aligned} 80 &=_2 1010000 & \qquad 80 &=_2 1010000 \\ 91 &=_2 1011011 & \qquad 68 &=_2 1000100 \\ 20 &=_2 0010100 & \qquad 20 &=_2 0010100 \end{aligned}$$

Finally, if Bob chooses to work with the third row, he must remove sixteen, then add eight, then remove four, then add two and then add one counter. Overall this amounts to $[80, 91, 20] \to [80, 91, 11]$.

$$\begin{aligned} 80 &=_2 1010000 & \qquad 80 &=_2 1010000 \\ 91 &=_2 1011011 & \qquad 91 &=_2 1011011 \\ 20 &=_2 0010100 & \qquad 11 &=_2 0001011 \end{aligned}$$

It should be clear that this approach is perfectly general except for one potential problem. Could it be that as Bob works along his chosen row, he is forced to add more counters than he removes? Fortunately this can never happen since the way Bob chooses his row ensures that the first thing he does is remove 2^k counters for some k. After that he only works with smaller powers of 2, so the greatest number of counters he could be required to return to the pile is $2^{k-1} + 2^{k-2} + \cdots + 2 + 1 = 2^k - 1$. This shows that overall Bob always removes at least one counter (as he did when working with the first row in the previous example).

The argument above is a full, if somewhat informal, proof of our characterisation of the $\mathcal{L}$-positions in three pile nim. We hope that presenting the proof by giving a recipe for winning moves will allow you to actually play nim perfectly against unsuspecting opponents.

Exercise 5d

1. Determine whether the following nim positions are $\mathcal{W}$ or $\mathcal{L}$ and find the $\mathcal{L}$-positions which follow the $\mathcal{W}$-positions.
   (a) $[15, 16, 17]$;
   (b) $[5, 10, 15]$;
   (c) $[16, 21, 25]$.

2. Investigate the $\mathcal{L}$-positions in nim with more than three piles.

3. Under the normal play convention the player who makes the last move wins. The opposite convention, where the player who makes the last move loses, is called the *misère* convention. Investigate the $\mathcal{L}$-positions when three pile nim is played with the misère convention.

4. The figure below shows a position in the game of *sliding*. The game is played with a long strip of square cells, some of which contain coins. A move consists of taking a coin and sliding a positive whole number of cells to the right, without the coin passing over or ending on top of another coin. The last player to move wins.
   Determine whether the position shown is a $\mathcal{W}$ or an $\mathcal{L}$-position and, if it is $\mathcal{W}$, determine the $\mathcal{L}$-positions which follow it.

5.6 Nim-values

Ever since discovering the labelling rules, Alice and Bob have become fairly expert at analysing individual impartial games. To keep things interesting they decide to try playing multiple games at once. Given two impartial games which they call X and Y, they define the *sum*, $X+Y$, to be the game where a turn consists of making a move in either X or Y but not both. The player who cannot complete a turn because they have no legal moves in either game loses.

Of course, $X+Y$ is an impartial game so can be analysed using the labelling rules. However, the game graph for $X+Y$ is often unmanageably large and complicated, and we can make a number of useful observations about $X+Y$ without thinking about its game graph.

In what follows we will often refer to games and their initial positions interchangeably. For example, we might say that a game X is $\mathcal{L}$, meaning that its initial position is an $\mathcal{L}$-position. We might also say that a game sum $X+Y$ is a $\mathcal{W}$-position, meaning that its initial position is.

Problem 5.13

Prove the following.

(a) If X and Y are both $\mathcal{L}$, then $X+Y$ is $\mathcal{L}$.

(b) If one of X and Y is $\mathcal{L}$ and the other is $\mathcal{W}$, then $X+Y$ is $\mathcal{W}$.

(c) If X and Y are both $\mathcal{W}$, then $X+Y$ may or may not be $\mathcal{L}$.

To prove (a) we imagine that Alice goes first in the combined game $X+Y$. If she moves in X then, since X is $\mathcal{L}$, Bob can respond so as to return X to an $\mathcal{L}$-position. He can do the same if Alice plays in Y so can never run out of moves before Alice does.

Statement (b) follows easily from (a). We suppose, without loss of generality, that X is $\mathcal{L}$ and Y is $\mathcal{W}$. This means if Alice moves first in $X+Y$ she can make a move in Y such that Y becomes an $\mathcal{L}$-position. Now (a) implies that Alice can force a win since Bob is faced with the sum of two $\mathcal{L}$-positions.

Statement (c) comes in two parts. First we exhibit two $\mathcal{W}$-positions whose sum is a $\mathcal{W}$-position. There are many examples. We can take X to be a game of nim with a single pile of two stones and Y to be a game of nim with a single pile of only one stone. These are both $\mathcal{W}$-positions but their sum is the nim position $[1,2]$ which is also a $\mathcal{W}$-position, since the first player can win by moving to $[1,1]$.

Finally we show that the sum of two $\mathcal{W}$-positions can be $\mathcal{L}$. Suppose X is any game and consider the sum $X+X$ composed of two identical copies of X. A simple symmetry argument shows that $X+X$ is an $\mathcal{L}$-position, since Bob can mimic Alice's moves. This is true regardless of whether X is $\mathcal{W}$ or $\mathcal{L}$, so the sum of two $\mathcal{W}$-positions can certainly be an $\mathcal{L}$-position.

Statement (c) naturally suggests that we investigate more closely when it is true that two $\mathcal{W}$-positions sum to an $\mathcal{L}$-position, and we will spend the rest of this section doing just that.

The observation that $X+X$ is always $\mathcal{L}$ motivates the following definition.

We call two games X and Y *equivalent* if, and only if, $X+Y$ is $\mathcal{L}$.

For example, the nim position $[1,2,0]$ is equivalent to the position $[0,0,3]$, or more succinctly, $[1,2] \equiv [0,3]$. Using table 5.3 we see that $[5,6]$ is also equivalent to $[0,3]$.

This raises an interesting question. Since $[1,2]$ and $[5,6]$ are both equivalent to the same game, $[0,3]$, does it follow that they are equivalent to each other? Put another way, does the fact that $[1,2,3]$ and $[5,6,3]$ are both $\mathcal{L}$ imply that $[1,2,5,6]$ is $\mathcal{L}$?

More generally we might ask whether two games which are equivalent to some third game are always equivalent to each other.

Problem 5.14

Prove that if game X is equivalent to game Y and game Y is equivalent to game Z, then game X is equivalent to game Z.

We must show that if Bob goes second, he can win $X+Z$ provided he can win $X+Y$ and $Y+Z$. Put another way, we must show that Bob can prevent Alice from making the last move in $X+Z$.

Our argument must use the fact that there is a game Y such that $X+Y$ and $Y+Z$ are both $\mathcal{L}$-positions. Indeed, since we make no assumptions about the nature of the three, possibly quite different, games X, Y and Z, there is essentially nothing else we can use.

Suppose that Alice makes a move in $X+Z$. We will show that Bob can respond to this move. Moreover, if we call the positions of X and Z after Bob's move $X*$ and $Z*$, we will show that Bob can find a game $Y*$ which is equivalent to both $X*$ and $Z*$. This is enough to ensure that Bob can force a win. He simply repeats the process after every move Alice makes, so is never left without a move.

Bob will use the game Y in an ingenious thought experiment. In the real world, he and Alice are playing two games, namely X and Z, but Bob imagines them playing four games: X, Z and two copies of Y. He thinks of this display as the sum $(X+Y)+(Y+Z)$. We will call a move which returns a $\mathcal{W}$-position to an $\mathcal{L}$-position a *good move*.

We suppose that Alice's first move changes the position of X to X' and denote this by $X \to X'$. Now since $(X+Y)$ was an $\mathcal{L}$-position, it must be that $(X'+Y)$ is $\mathcal{W}$, so Bob has a good move in this sum. If there is a good move of the form $X' \to X''$, then we are done. Bob can make the move $X' \to X''$ in the real world and set $X* = X'', Y* = Y$ and $Z* = Z$.

If, on the other hand, the only good move in $(X'+Y)$ is of the form $Y \to Y'$, then the thought experiment continues. Bob imagines making this move, and also imagines that Alice makes the same move in the sum $(Y+Z)$. We now have $(X'+Y')+(Y'+Z)$ with Bob to play.

As before, there must be a good move in $(Y'+Z)$. If there is a good move $Z \to Z'$ we are done: Bob can make this move in the real sum $(X'+Z)$ and set $X* = X', Y* = Y', Z* = Z'$.

However, if the only good moves in $(Y'+Z)$ are of the form $Y' \to Y''$ then the thought experiment continues yet again. Bob imagines making that move, and imagines Alice mimicking him in the sum $(X'+Y')$. Now we have $(X'+Y'')+(Y''+Z)$ with Bob to play. Again there are two cases. If Bob can make a good move $X' \to X''$ in $(X'+Y'')$, he can do so in the real sum $(X'+Z)$ and set $X* = X'', Y* = Y'', Z* = Z$. If his only good move in $(X'+Y'')$ is in Y'' he should make this imaginary move, imagine Alice copying him, and continue the thought experiment.

Since Y is finite, the sequence of imaginary moves cannot go on forever. Eventually the thought experiment will end and Bob will be able to make a move in the real sum. Once this happens the positions will be $X*$ and $Z*$ and there will be a game $Y*$ which is equivalent to each of these. This ensures that Bob can always find a response to Alice's next move, and is therefore bound to win in the end.

This result is tremendously important since it opens up a very general approach for analysing sums of games. To determine whether or not

$G_1 + H_1$ is an $\mathcal{L}$-position, we need to find out whether G_1 and H_1 are equivalent. This may be hard to do directly. However, we may be able to simplify matters by finding other games G_2 and H_2 which are equivalent to G_1 and H_1 respectively, and then asking whether or not G_2 and H_2 are equivalent. It would be particularly delightful if we could take G_2 and H_2 to be nim positions with a single pile of counters. In this case G_1 and H_1 are equivalent precisely when G_2 and H_2 contain the same number of counters. Of course, a cynic might suspect that such a delightfully straightforward scenario would be a rarity, but the following result shows that the opposite is true.

Problem 5.15

Prove that every finite impartial game is equivalent to a single nim pile.

We start by studying a simple example of a game G, whose game graph has seven vertices labelled a to g as shown in figure 5.12. If every impartial game really is equivalent to a nim pile, then it should be possible to label each vertex with the size of the single nim pile which is equivalent to starting the game at that vertex. These labels are called the *nim-values* of the positions. The vertices have been labelled with some non-negative integers which will (miraculously) turn out to be the nim-values of the positions.

The labelling rules allow us to see that vertices e and g are $\mathcal{L}$-positions, and thus equivalent to piles of zero counters. However, they do not allow us to determine which nim pile is equivalent to each of the $\mathcal{W}$-positions. To prove that the labels in figure 5.12 are correct, we need to look at the game graph of $(G + [3])$, the sum of G and a nim pile with three counters.

We refer to positions in this game sum using the ordered pairs $(a, 3)$, $(b, 3)$, $\ldots$, $(g, 3)$, $(a, 2)$, $\ldots$, $(g, 0)$ where the letter corresponds to the position in G and the number corresponds to the size of the nim pile. We call the game graph of G shown in figure 5.12 the *small graph*, and the game graph of $G + [3]$ shown in figure 5.13 the *large graph*, and note that the large graph consists of four copies of the small graph stacked one above the other.

A move in G corresponds to staying on the same level in the large graph by following a black arrow, while a move in $[3]$ corresponds to

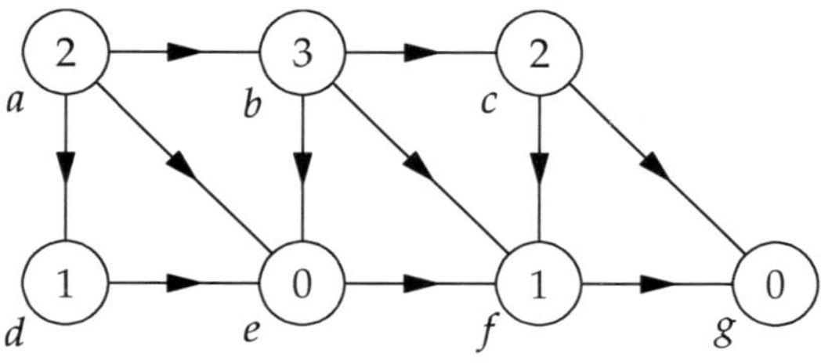

Figure 5.12

moving from a vertex to *any* vertex directly below it by following *one or more* grey double arrows.

We can now use the labelling rules to study the large graph. It will be helpful to refer to the positions $(g, 0), \ldots, (g, 3)$ collectively as the g positions and so on.

The position $(g, 0)$ has no followers and is therefore $\mathcal{L}$. This justifies the 0 label on g in the small graph and implies that $(c, 0)$, $(f, 0)$ and the three remaining g positions are all $\mathcal{W}$.

Now we turn our attention to the f positions since these are followed either by other f positions or by g positions, and every g position is already labelled.

The position $(f, 1)$ is followed by $(f, 0)$ and $(g, 1)$, both of which are $\mathcal{W}$, so $(f, 1)$ is $\mathcal{L}$. This justifies the 1 label in the small graph and implies that $(b, 1), (c, 1), (e, 1)$ and the remaining f positions are all $\mathcal{W}$.

Next we look at the c positions, since these are either followed by other c positions, or by f or g positions which are already labelled. The fact that $(c, 0)$ and $(c, 1)$ are followed by $\mathcal{L}$-positions corresponds exactly to the fact that, in the small graph, c is followed by vertices labelled 0 and 1. The next position to consider is $(c, 2)$. The fact that no follower of c in the small graph has a nim-value of 2 means that $(c, 2)$ is not followed by an $\mathcal{L}$-position on the 2 level. Moreover, $(c, 2)$ is the lowest c position we have not already established is $\mathcal{W}$, so the c positions which follow it are certainly $\mathcal{W}$. We conclude that $(c, 2)$ is $\mathcal{L}$, and that the label 2 on vertex c in the small graph is correct.

A similar argument can be applied to any position p in any given game G provided all its followers have already been labelled in the small graph. Suppose that the set of labels on followers of p is $Z = \{z_1, z_2, \ldots\}$.

We work up the p positions $(p, 0), (p, 1), \ldots$ looking for an $\mathcal{L}$-position.

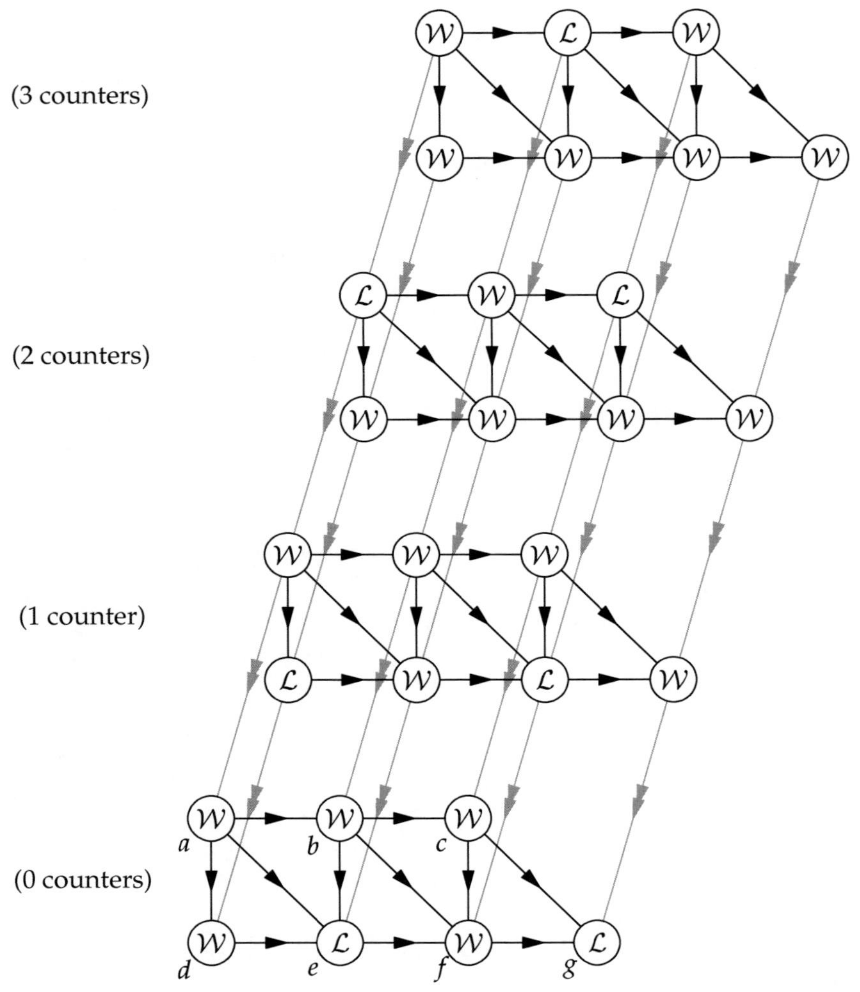

Figure 5.13

If $0 \notin Z$, then $(p, 0)$ has no $\mathcal{L}$-positions among its followers. In this case we may label p with a 0 in the small graph.

If $0 \in Z$ and $1 \notin Z$, then $(p, 0)$ has an $\mathcal{L}$-position among its followers on the 0 level, so is $\mathcal{W}$, but $(p, 1)$ has no $\mathcal{L}$-positions among its followers so is $\mathcal{L}$. In this case we label p with a 1.

Similarly $(p, 2)$ is $\mathcal{L}$ precisely if $0 \in Z, 1 \in Z$ and $2 \notin Z$. In general we see that p in the small graph receives the label $mex(Z)$.

You should take a moment to check that the other labels shown in figure 5.12 are indeed correct, and also to think about how this result relates to the way in which we populated table 5.3 in our analysis of nim.

We can summarise what we have just learned in a single *advanced labelling rule*:

ADVANCED LABELLING RULE

(i) We label positions with their *nim-values*.

(ii) The nim-value of a position is the *mex* of the nim-values of its followers.

The argument used in problem 5.3 can be easily adapted to show that every position in a finite impartial game can be assigned a nim-value using this rule. The $\mathcal{L}$-positions are labelled zero while the $\mathcal{W}$-positions receive positive labels.

The major strength of this new rule is the fact that two positions sum to an $\mathcal{L}$-position if and only if they have the same nim-value. We have worked hard to discover the advanced labelling rule, and the *mex* function it relies on is not particularly familiar. With hindsight however, it is easy to see how well suited it is to determining when the sum of two $\mathcal{W}$-positions is $\mathcal{L}$.

Suppose that the starting positions in games X and Y both have nim-value n. We claim that Bob can win the game sum $X + Y$ by ensuring the nim-values are always equal at the end of his turn. Assume Alice moves in X taking it to a position with nim-value m. If $m < n$ then Bob can move in Y to a position with nim-value m since, by the advanced labelling rule, every number less than n is the nim-value of a follower of Y's start position. If, on the other hand, $m > n$ then Bob can move in A such that the nim-value of the position returns to n. This also follows from the advanced labelling rule since m is the *mex* of the values of the position's followers.

Position	Followers	Nim-values	Nim-value of position
(1)			$mex(\emptyset) = 0$
(2)			$mex(\emptyset) = 0$
(3)	(1,2)	0	$mex(\{0\}) = 1$
(4)	(1,3)	1	$mex(\{1\}) = 0$
(5)	(1,4), (2,3)	0, 1	$mex(\{0,1\}) = 2$
(6)	(1,5), (2,4)	2, 0	$mex(\{0,2\}) = 1$

Table 5.5

Studying sums of games may seem like a rather artificial activity, given that most people tend to play one game at a time. However, we have already met a number of games which naturally break down into sums of smaller games as they are played. An archetypal example is Grundy's game, where players take it in turns to choose a pile of counters and divide it into two unequal piles. In problem 5.2 we used a game graph to show that the position with a single pile of eight counters is $\mathcal{W}$. Our new techniques will allow us to analyse the game rather more efficiently.

Problem 5.16

Find the nim-value of a pile of n counters in Grundy's game for every $n \leq 12$.

Recall from problem 5.2 on page 147 that we denote positions in Grundy's game using normal brackets, so $(1,1,3)$ means two piles of one counter and a pile of three counters.

Our aim is to build up a table of the nim-values for the games (n) where $n \leq 12$. We will build up the table one row at a time. For example, suppose we want to find the nim-value of the game (6). This value is the *mex* of the nim-values of the positions which follow (6). The position (6) is followed by $(1,5)$ and $(2,4)$ and, since piles of one or two counters cannot be touched in subsequent moves, these are equivalent to (5) and (4) respectively. If we assume that we have already found the nim-values of (5) and (4), then the nim-value of (6) is simply the *mex* of these two values.

This type of argument quickly allows us to build up table 5.5.

Start	Nim-values of followers	Nim-value of position
(1)		$mex(\emptyset) = 0$
(2)		$mex(\emptyset) = 0$
(3)	0	$mex(\{0\}) = 1$
(4)	1	$mex(\{1\}) = 0$
(5)	0, 1	$mex(\{0,1\}) = 2$
(6)	2, 0	$mex(\{0,2\}) = 1$
(7)	1, 2, 1	$mex(\{1,2\}) = 0$
(8)	0, 1, 3	$mex(\{0,1,3\}) = 2$
(9)	2, 0, 0, 2	$mex(\{0,2\}) = 1$
(10)	1, 2, 1, 1	$mex(\{1,2\}) = 0$
(11)	0, 1, 3, 0, 3	$mex(\{0,1,3\}) = 2$
(12)	2, 0, 0, 2, 2	$mex(\{0,2\}) = 1$

Table 5.6

The position (7) is more interesting. Its followers are $(1,6), (2,5)$ and $(3,4)$ and the first two are equivalent to (6) and (5) respectively. However, $(3,4)$ is a little more subtle. We may think of $(3,4)$ as the game sum $(3) + (4)$. Since the nim-values of (3) and (4) are 1 and 0 respectively, we see that the position $(3,4)$ in Grundy's game is equivalent to the position $[1,0]$ in nim, which clearly has nim-value 1.

Similarly we see that the position $(3,5)$ is equivalent to the nim position $[1,2]$ so the nim-value of $(3,5)$ is 3.

In general we may use our understanding of nim to compute the nim-value of any position (p,q). We look up the nim-values of the games (p) and (q) and use the nim addition table 5.3 to find the value of the sum $(p) + (q)$.

Table 5.6 shows the results of this process for the next few values of n thereby solving the problem. It also provides a second solution to problem 5.2 since (8) has nim-value 2 (not 0) and is therefore a $\mathcal{W}$-position
.

It is also worth noting that, for $n > 2$, the sequence of nim-values seems to repeat with period three. Unfortunately the nim-value of (13) turns out to be 3 rather than 0 so the pattern does not continue. It is conjectured that the sequence of nim-values for Grundy's game does eventually become periodic, but, at the time of writing, the question remains unresolved.

In problem 5.7 on page 154 we met the game of Kayles, which is played with a row of skittles. Players take it in turns to knock over either one skittle or two adjacent skittles. A simple symmetry argument showed that for a single row of skittles the first player can always win.

Problem 5.17

Alice and Bob are about to play a game of Kayles with twelve skittles when the fourth skittle spontaneously falls over leaving a group of three skittles and a separate group of eight skittles. Which player has a winning strategy now?

We can adapt both the method and the notation used in the previous problem. We use normal brackets to denote the lengths of consecutive blocks of skittles in the position. The spontaneous collapse of the skittle turned the boring position (12) into the more interesting position $(3,8)$ which is the same as the game sum $(3)+(8)$.

Now we can construct a table of nim-values for (n) where $n \leq 8$. The followers of (n) are all of the form (a,b) where $a+b$ equals $n-1$ or $n-2$ depending on whether one or two skittles were knocked over that turn. To find the nim-value of (a,b) we look up the nim-values of (a) and (b) and find the value of the corresponding two pile nim game. This can be done using table 5.3 or by writing the values in binary, and finding the binary number which makes the number of 1s in each column even.

The results are show in table 5.7. We note that all the nim-values after (0) are strictly positive, which confirms that a game consisting of a single row of skittles is a first player win. To solve the problem at hand, we need to find the nim-value of $(3,8)$. The nim-values of (3) and (8) are 3 and 1 respectively so the position has a nim-value of 2 and is therefore a first player win. To be sure of winning, Alice might move to (1,8) since (1) is the unique follower of (3) with a nim-value of 1. Alternatively, she could move to (3,6) or (3,2,4) since (6) and (2,4) both have nim-value 3.

Start	Followers	Nim-values	Nim-value of position
(0)			$mex(\emptyset) = 0$
(1)	(0)	0	$mex(\{0\}) = 1$
(2)	(0), (1)	0, 1	$mex(\{0,1\}) = 2$
(3)	(1), (2), (1,1)	1, 2, 0	$mex(\{0,1,2\}) = 3$
(4)	(2), (1,1), (3), (1,2)	2, 0, 3, 3	$mex(\{0,2,3\}) = 1$
(5)	(3), (1,2) (4), (1,3), (2,2)	3, 3 1, 2, 0	$mex(\{0,1,2,3\}) = 4$
(6)	(4), (1,3), (2,2) (5), (1,4), (2,3)	1, 2, 0 4, 0, 1	$mex(\{0,1,2,4\}) = 3$
(7)	(5), (1,4), (2,3) (6), (1,5), (2,4), (3,3)	4, 0, 1 3, 5, 3, 0	$mex(\{0,1,3,4,5\}) = 2$
(8)	(6), (1,5), (2,4), (3,3) (7), (1,6), (2,5), (3,4)	3, 5, 3, 0 2, 2, 6, 2	$mex(\{0,2,3,5,6\}) = 1$

Table 5.7

5.7 What next?

There is a sense in which the theory of nim-values is the end of the story for impartial games, since they provide a systematic way to analyse not only individual games, but sums of games as well. However, many interesting questions still remain.

In our analysis of Grundy's game in problem 5.16, we mentioned that, while individual nim-values are easy to compute, the long term behaviour of the sequence of nim-values is not yet well understood.

There is also the question of how to classify games and invent new ones. For example, the games Nim, Kayles and the game in 2 of exercise 5e are all examples of *take and break* games. In these games a position is a set of piles of counters, and a move consists of taking some counters from a pile, or breaking the pile into smaller piles or both. By varying the number of counters which may be removed, or the number of piles a pile may be broken into, we can construct any number of related games. Some of these games are easy to analyse, but a full theory of take and break games is still some way off.

Other games are hard to pin down purely because the set of positions is too vast and unruly to study directly. For example, despite the fact that the game of cram is essentially trivial if at least one side of the board is even, cram on a 9×9 board can still be played competitively.

There are even impartial games where simply finding good notation for the positions, and determining which follow which, is a significant challenge.

The starting position in a game of *sprouts* consists of a number of dots on a piece of paper. A move consists of choosing two dots, joining them with a (possibly curved) line which does not cross any previously drawn lines, and adding a dot half way along the new line. It is also possible to join a dot to itself. The only other condition on moves is that the number of line segments emanating from any dot may never exceed three.

Figure 5.14 shows two possible games starting with just two dots, one lasting four moves and the other lasting five. For larger number of dots, the topological nature of the condition that lines should not cross makes systematically listing legal positions a very delicate task. It is conjectured that the first player has a winning strategy if the number of dots is congruent to 3, 4 or 5 modulo 6, but a proof of this claim is still elusive.

Finally, impartial games behave very differently if we abandon the normal play convention in favour of the *misère* convention that the last player to make a move loses. In general misère games are more resistant to analysis than their normal play counterparts, and the theory of misère games, while extensive, is less complete than the theory of nim-values. (It turns out that the theory of misère nim is not much harder than that of normal nim, but this is unusual.)

In this chapter we have also mentioned a handful of games drawn from mathematical competitions. Games in competitions tend to rely on the key principles of symmetry and invariance, and often include elements of basic counting and number theory. They are not always impartial, but never require knowledge of the general theory of partisan games. However, readers interested in partisan games should rest assured that these games possess a rich and rapidly developing theory.

The first reference for mathematical games is undoubtedly [7]. Many of the games from this chapter are discussed there in greater depth and generality.

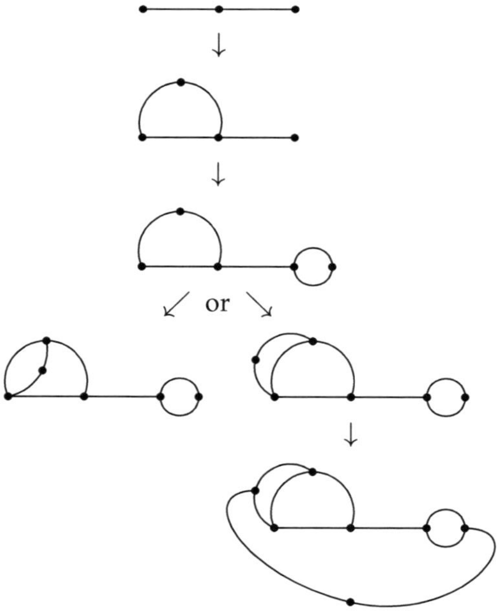

Figure 5.14

Exercise 5e

1. In problem 5.17 we studied the sequence $k_0, k_1, k_2, \ldots$ where k_n is the nim-value of a position (n) in Kayles.
The first 192 terms of this sequence are as follows:
0, 1, 2, 3, 1, 4, 3, 2, 1, 4, 2, 6, 4, 1, 2, 7, 1, 4, 3, 2, 1, 4, 6, 7,
4, 1, 2, 8, 5, 4, 7, 2, 1, 8, 6, 7, 4, 1, 2, 3, 1, 4, 7, 2, 1, 8, 2, 7,
4, 1, 2, 8, 1, 4, 7, 2, 1, 4, 2, 7, 4, 1, 2, 8, 1, 4, 7, 2, 1, 8, 6, 7,
4, 1, 2, 8, 1, 4, 7, 2, 1, 8, 2, 7, 4, 1, 2, 8, 1, 4, 7, 2, 1, 8, 2, 7,
4, 1, 2, 8, 1, 4, 7, 2, 1, 8, 2, 7, 4, 1, 2, 8, 1, 4, 7, 2, 1, 8, 2, 7,
4, 1, 2, 8, 1, 4, 7, 2, 1, 8, 2, 7, 4, 1, 2, 8, 1, 4, 7, 2, 1, 8, 2, 7,
4, 1, 2, 8, 1, 4, 7, 2, 1, 8, 2, 7, 4, 1, 2, 8, 1, 4, 7, 2, 1, 8, 2, 7,
4, 1, 2, 8, 1, 4, 7, 2, 1, 8, 2, 7, 4, 1, 2, 8, 1, 4, 7, 2, 1, 8, 2, 7...
Predict the next term of the sequence, and prove that your prediction is correct.

2. Consider the variant of Kayles where the only legal move is to knock over two adjacent skittles. (So the position (1) has no followers.)
Find $a_0, a_1, \ldots, a_{11}$ where a_n is the nim-value of the position (n) in this game.

3. Consider the variant of Kayles where the only legal move is to knock over a single skittle which is directly between two other skittles. (So the position (2) has no followers.)
Find $b_0, b_1, \ldots, b_8$ where b_n is the nim-value of the position (n) in this game.
Without further calculation, predict the values of b_9, b_{10}, b_{11} and explain why your prediction is correct.

4. In 1935 T. R. Dawson proposed the following chess puzzle.
Suppose that a chessboard with three rows and n columns has a white pawn on every cell in bottom row and a black pawn on every cell in the top row. (The set up for $n = 8$ is shown in figure 5.15.)
As in normal chess, the pawns may move forward into an empty space, or may move diagonally to capture an opposing piece. Unlike normal chess, we add the rule that a player who is able to capture an opponent's piece must do so.

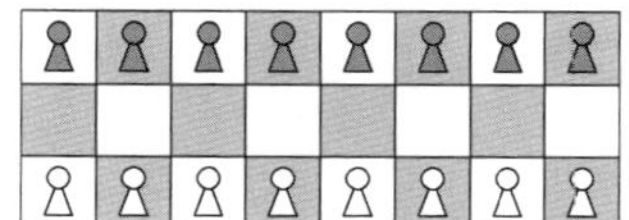

Figure 5.15

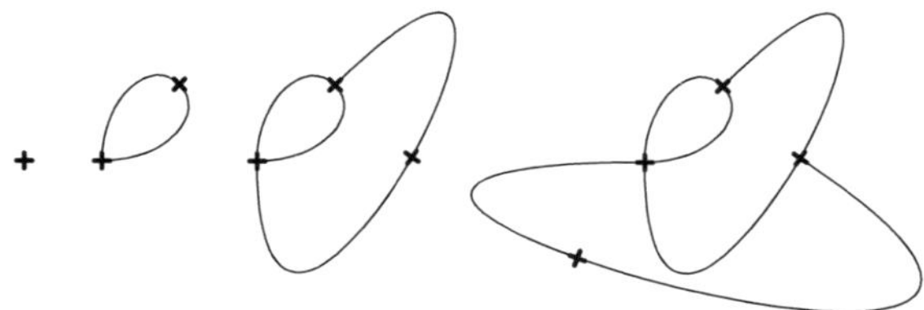

Figure 5.16

White moves first, and the last player who is able to make a move wins (normal play).

(a) Who wins the game for $n = 1, 2, 3, 4$?

(b) Explain why this game is equivalent to an impartial game, even though each player can only move their own pawns.

(c) Find the nim-values for this impartial game for $n = 1, 2, \ldots, 7$.

(d) Predict the next three nim-values and prove that your prediction is correct.

5. The game of *Brussels sprouts* is a (less difficult) variant of sprouts. The game starts with a number of crosses drawn on a piece of paper. If we think of a cross as a dot with four free ends, then a move consists of joining two free ends with a (possibly curved) line and then drawing a short stroke across the new line to create two new free ends.
A game with only one starting cross is shown in figure 5.16.
If Alice and Bob play with just two crosses to start with, who wins? Can you generalise this to n crosses?

Chapter 6

Ramsey Theory

In his short life, Frank P. Ramsey made lasting contributions to economics, epistemology and logic. He also, somewhat accidentally, founded a branch of combinatorics. Ramsey's theorem, which we shall meet in various forms throughout the chapter, was published in a paper entitled 'On a problem in formal logic' ([6]). The theorem was stated, proved and discussed in the first few pages of the fairly lengthy paper, and was certainly not the main result in Ramsey's sights, yet this charming result inspired countless related problems, many of which remain unresolved.

6.1 Order in disorder

Ramsey theory is often described as asking:

Given a large amount of disorder, is it always possible to find a small amount of order within it?

Problem 6.1

Let G be the set of points (x, y) in the plane such that x and y are integers in the range $1 \leq x, y \leq 2011$. A subset S of G is said to be parallelogram-free if there is no proper parallelogram with all its vertices in S. Determine the largest possible size of a parallelogram-free subset of G. *(A parallelogram is called* proper *if it does not have all its vertices on a line.)* [British Mathematical Olympiad Part Two 2011]

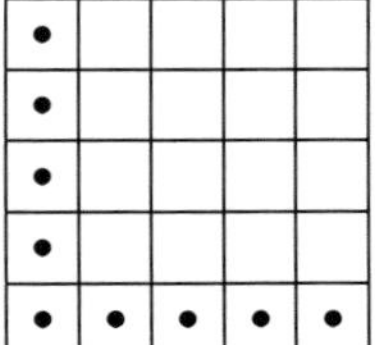

Figure 6.1

It seems likely that the number 2011 is not particularly significant, so we should be able to solve the problem for an $n \times n$ grid of points.

To get a feel for the problem we start by trying to construct some large parallelogram-free sets. Choosing all the points in the bottom row and leftmost column, as shown in figure 6.1, gives a set of size $2n - 1$ which is clearly parallelogram-free.

We cannot add any points to this parallelogram-free set without forming a rectangle, but this does not prove that no larger set exists. For that we must begin by assuming that we have *any* set with $2n$ points and show that it contains a parallelogram.

Since we may not make any assumptions about these $2n$ points, they represent the disorder in the problem. The order we seek is the parallelogram.

The $2n$ is suggestive. If there are $2n$ points, there are, on average, two points per row. We might, therefore, ask whether we can solve the special case where every row contains exactly two points. In this case we consider the distance between the points in each row. There are n rows, but only $n - 1$ available distances, so, by the pigeonhole principle, there is a distance which occurs in more than one row. Taking the points in two of these rows gives the required parallelogram.

To adapt this argument to the general case we say that row i contains k_i points. For each non-empty row we consider the $k_i - 1$ distances from the leftmost point in the row to the other points in the row. If two of these distances are the same we are done since the two equal distances cannot occur in the same row. All the distances are at most $n - 1$ so it remains to check that we have considered at least n distances. If we allow i to range over the non-empty rows, then the number of distances considered is $\sum_i (k_i - 1) = \sum_i k_i - \sum_i 1$. The first sum is exactly $2n$ and the second counts the number of non-empty rows which is at most n. Thus the number of distances we have considered is at least n as required.

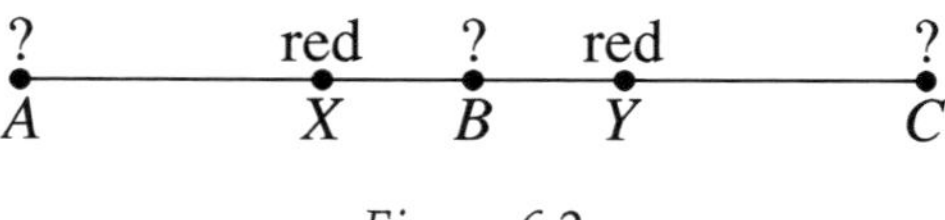

Figure 6.2

Thus, the largest possible size of a parallelogram-free subset of a 2011×2011 grid is 4021.

It should not surprise us that the pigeonhole principle played a crucial role in this solution. The pigeonhole principle is all about finding order given enough disorder, and it underpins much of Ramsey theory.

The next two problems involve assigning colours to points and seeking a particular configuration of points, all of which are the same colour. Such configurations are called *monochromatic*.

Problem 6.2

Suppose that every point on a straight line is coloured either red or blue. Prove that it is possible to find three evenly spaced points which are the same colour.

We start with two points X and Y that are the same colour. We may assume that they are red. Now let B be the midpoint of XY, let A be such that $AX = XY$ and C be such that $XY = YC$. This makes X the midpoint of AY and Y the midpoint of XC as shown in figure 6.2. If any of A, B or C are red, then we have three evenly spaced red points, but if none of them are red, then they form a set of three evenly spaced blue points.

We note that our solution involved focusing on a small finite number of the available points on the line. We will use a similar approach in the next problem.

Problem 6.3

Suppose that every point in the plane is coloured red or blue. Prove that it is possible to find a monochromatic rectangle.

Let us consider a finite grid of points.

If our grid of points has three rows, then every column contains at least two points that are the same colour. If, say, the first and third points are red, then finding a red rectangle is equivalent to finding another column whose first and third points are red. In other words, we need a second column whose colour pattern is almost the same as that of the first column. Now we can solve the problem by asking for more than we need: we can insist that two columns have identical colour patterns.

There are eight ways to colour a column of three points with two colours. So, if our grid has nine columns, two of the columns have identical colour patterns. These two columns contain the corners of a monochromatic rectangle.

The next few problems all involve a group of people at a party. We assume that any two guests are either friends or strangers, and that there is no other restriction on the relationships that might exist. For example, we do not require that each of the guests should have at least one friend at the party. In fact, we can regard the relationships as being completely disordered. Within this disorder, the order we seek is a small group of people, all of whom have the same relationship to each other. In other words, they are mutual friends or mutual strangers.

We begin with the simplest possible problem of this type.

Problem 6.4

For which values of n can we guarantee that at a party with n guests there are either three mutual friends or three mutual strangers?

It seems plausible that a party with, say, twenty guests will always have sizeable groups of mutual friends or mutual strangers. However, there are many millions of ways in which the relationships between twenty people might be arranged, so it is also conceivable that a particularly devious host might be able to construct a party without such groups. It is even conceivable that there are no values of n which guarantee three friends or three strangers.

To aid our discussion of this problem, we will rephrase it in the language of graph theory, as introduced in section 2.2 of chapter 2. We represent the party as a graph where the guests are the vertices. Since we are looking for groups of friends or groups of strangers, we will use two different types of edge. We join two people with a red edge if they are friends and a blue edge if they are strangers. We are then looking for a triangle which is completely red or completely blue. We will call such a

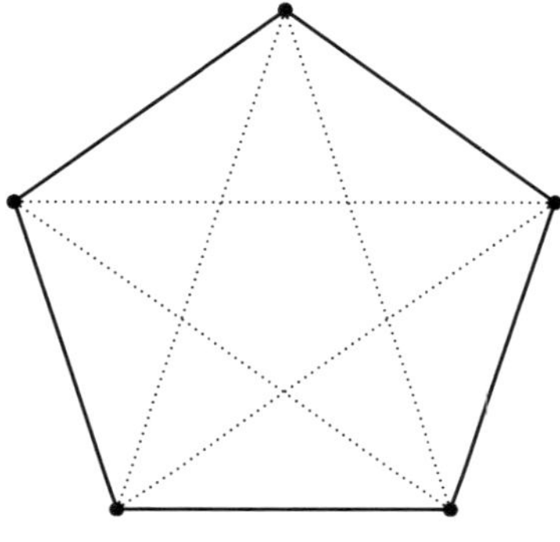

Figure 6.3

triangle *monochromatic*, noting that we are now talking about the colours of the edges not the vertices.

We begin with an explicit colouring of a pentagon which does not contain a monochromatic triangle. This will show that $n > 5$. One such colouring is shown in figure 6.3 with solid lines representing red edges and dotted lines representing blue edges.

Having shown that $n > 5$, our next task is to consider $n = 6$. If we try to find a triangle-free colouring of a hexagon, we run into difficulties. The problem is that if two edges from a vertex v, say $\{v, x\}$ and $\{v, y\}$, are both red, then the edge between x and y must be blue to avoid a red triangle. This is fine, but if another edge containing v, say $\{v, z\}$, is also red, then all the edges between the vertices x, y and z are forced to be blue, which gives us a blue monochromatic triangle. The only way out is if only two edges from v are red. However, if v is contained in at most two red edges, then it must be in at least three blue edges, which leads to the same difficulties, but with the colours reversed.

This rather wordy argument can be tidied up as follows.

Consider any vertex v. It is contained in five edges, so, by the pigeonhole principle, at least three edges are the same colour. We may call them $\{v, x\}, \{v, y\}$ and $\{v, z\}$ and may assume, without loss of generality, that they are red. Now consider the edges $\{x, y\}, \{y, z\}$ and $\{z, x\}$. If any one of them is red we have a red triangle, and otherwise we have a blue triangle. The situation is shown in figure 6.4.

This proof, together with the triangle-free colouring of the edges of the pentagon, shows that the solution to problem 6.4 is $n \geq 6$.

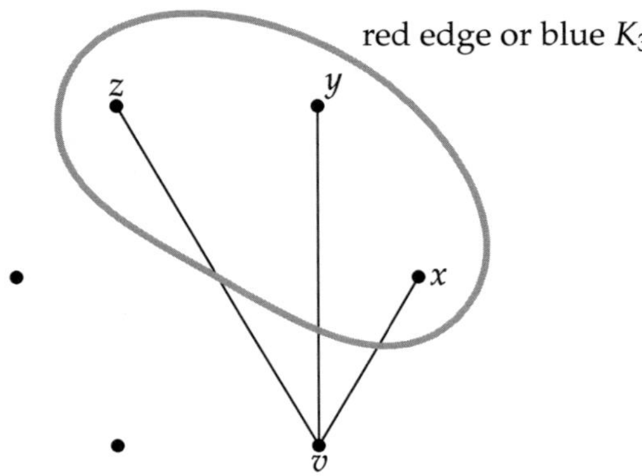

Figure 6.4

Exercise 6a

1. Suppose that the points in the plane are coloured with k different colours. More concisely we might say that the plane is *k-coloured*. Prove that it is possible to find a monochromatic rectangle.

2. Problem 6.3 showed that if a 3×9 grid is 2-coloured, then there is always a monochromatic rectangle. Show that this is also true when a 3×7 grid is 2-coloured, but not necessarily true when a 3×6 grid is 2-coloured.

3. Prove that if every point in the plane is coloured red or blue, then there is an equilateral triangle whose vertices are all the same colour. (*Hint: start with the result from problem 6.2.*)

4. Describe a way of colouring the plane with two colours such that no equilateral triangle with side length 1 unit has all three vertices the same colour.

5. Suppose that every point in the plane is coloured red or blue. Prove that it is possible to find ten lines parallel to the x-axis and ten lines parallel to the y-axis such that the hundred points of intersection are all the same colour.

6. Seventeen people correspond by mail with one another - each one with all the rest. In their letters only three different topics are discussed. Each pair of correspondents deals with only one of these topics. Prove that there are at least three people who write to each other about the same topic. [International Mathematical Olympiad 1964]
(*Hint: this is similar to problem 6.4 but with three kinds of relationship.*)

6.2 Ramsey numbers

In problem 6.4 we established that every party with $n \geq 6$ guests has either three mutual friends or three mutual strangers. Using some terminology from chapter 2 we might rephrase this as follows.

Whenever the edges of the complete graph K_6 are 2-coloured, there is a monochromatic K_3.

At this point it is natural to ask which values of n guarantee that at a party with n guests there are either four mutual friends or four mutual strangers. Equivalently, we might ask which n ensure that whenever the edges of K_n are 2-coloured there is a monochromatic K_4.

This problem turns out to be considerably harder than the previous one. Before tackling it we need to look extremely carefully at our solution to problem 6.4 and ask ourselves both what steps we took and why they worked. We started with an arbitrary vertex v and focused on a set of three vertices which were joined to v by red edges. We then made use of the following property of the number three.

Any set of three vertices in our graph contains a red K_2 (that is an edge) or a blue K_3.

This fact allowed us to establish a more interesting property of the number six, namely,

Any set of six vertices in our graph contains a red K_3 or a blue K_3.

The two properties are clearly very similar. The crucial difference is that the first is less symmetrical. This makes it less appealing, but also suggests that it might be worthwhile to ask the following question.

How large must a 2-coloured complete graph be to ensure the existence of a red K_3 or a blue K_4?

Returning to the original interpretation and generalising slightly, we might ask:

For which values of n is it the case that every party with n guests has a group of s mutual friends or a group of t mutual strangers?

More formally, we make the following definition.

The *Ramsey number* $R(s,t)$ is the smallest n such that, whenever the edges of K_n are coloured red or blue, there is either a red K_s or a blue K_t.

Thus, for example, problem 6.4 showed that $R(3,3) = 6$ while its solution made use of the fact that $R(2,3) = 3$. Our current aim is to find the value of $R(4,4)$, but our new definition allows us to consider a problem which is harder than finding $R(3,3)$ but easier than finding $R(4,4)$.

Problem 6.5

Find an upper bound for $R(3,4)$.

The problem asks us to find an n such that every party with n guests has three friends or four strangers. Before we solve it, let us gather together a few facts about Ramsey numbers.

- $R(2,t) = t$;
- $R(s,t) = R(t,s)$;
- $R(3,3) = 6$.

The first fact is merely the observation that if we colour the edges of K_t red or blue, then we either use a red edge somewhere or have a totally blue K_t.

The second fact is also evidently true. If a party with n guests always has s friends or t strangers, then a party with n guests always has s strangers or t friends. This is clear because we can assign the colours red and blue to the types of relationship either way round.

Now we are ready for our first solution to problem 6.5. We claim that $R(3,4) \leq 12$. We suppose that the edges of K_{12} are coloured red or blue and choose any vertex v in the graph. We will call another vertex w a *red-neighbour* or *blue-neighbour* of v according to the colour of the edge $\{v,w\}$.

The vertex v has eleven neighbours in total so either it has at least six blue-neighbours or at least six red-neighbours.

In the first case we focus on a K_6 whose vertices are all blue-neighbours of v. Since $R(3,3) = 6$, this K_6 either contains a red K_3 or a blue K_3. If there is a red K_3 we are done and if there is a blue K_3 we can add v to it to form a blue K_4 as required.

In the second case we focus on a K_6 whose vertices are all red-neighbours of v. If this K_6 contains a red edge, then we may add v to this edge to obtain a red K_3. If the K_6 contains no red edges, then it certainly contains a blue K_4 and we are done.

This is a successful solution to the problem, but the fact that in the last case we had a blue K_6 when we only need a K_4 suggests we may be able to do better. In fact we can adapt the proof to obtain the bound $R(3,4) \leq 10$.

Suppose the edges of K_{10} are coloured red or blue and choose a vertex v. This vertex has nine neighbours. Since $9 > 3 + 5$ we conclude that v either has four red-neighbours or six blue-neighbours.

We consider these two possibilities in turn. The case where v has four red-neighbours is shown on the left in figure 6.5. We note that, since $R(2,4) = 4$, the red-neighbours of v either contain a blue K_4 or a red edge. In the first case we are done and in the second we add v to the red K_2 to form a red K_3 as required.

The case where v has six blue-neighbours is shown on the right in figure 6.5. Since $R(3,3) = 6$, these six vertices include a monochromatic K_3. If it is red we are done. If it is blue we add v to form a blue K_4.

Problem 6.5 only asked us for an upper bound for $R(3,4)$ while problem 6.4 demanded the exact value of $R(3,3)$. Finding the exact value involved a cunning argument to prove that $R(3,3) \leq 6$ and an explicit triangle-free colouring of K_5 to show that $R(3,3) > 5$. The next problem gives a lower bound for $R(3,4)$.

Problem 6.6

Prove that $R(3,4) \geq 8$.

The problem requires us to find a colouring of the edges of K_7 which does not contain a red triangle or a blue K_4. Fortunately this is reasonably straightforward. We colour a 7-cycle red and all other edges blue as shown in figure 6.6. Clearly there is no red triangle in this colouring. To check

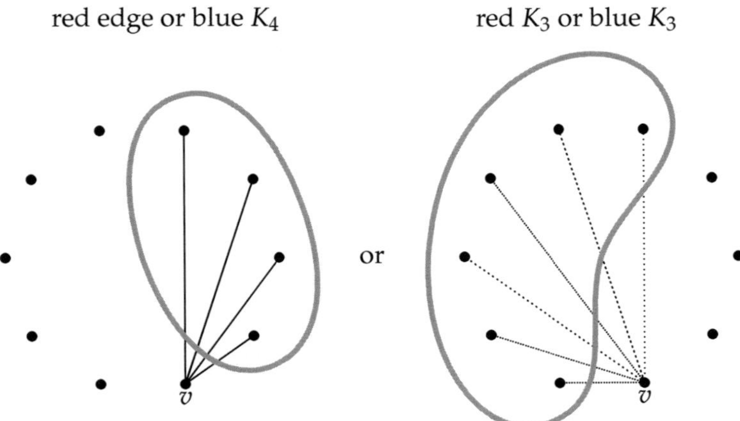

Figure 6.5

that we have no blue K_4 we observe that it is impossible to choose four out of the seven vertices without having two which are adjacent to each other in the red 7-cycle. These two vertices are joined by a red edge and so cannot be part of a blue K_4.

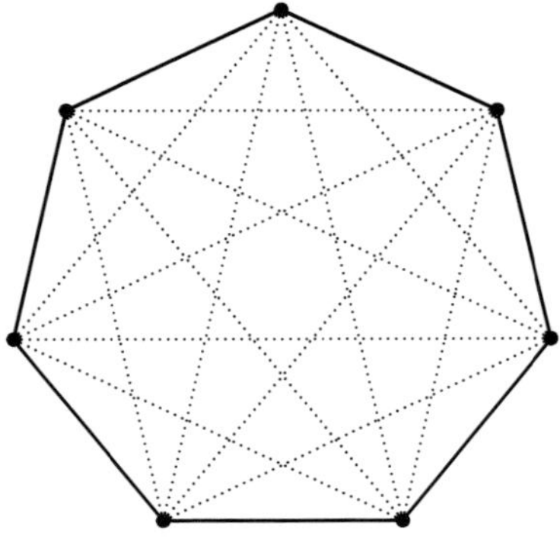

Figure 6.6

It is possible to describe this colouring in a way which we will be able to generalise when we come to colour larger graphs. We suppose that the vertices of our K_7 are the vertices of a regular heptagon with side

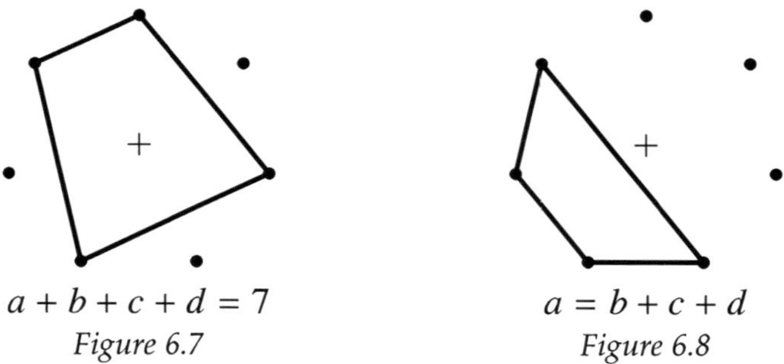

Figure 6.7

Figure 6.8

length one, and that the vertices are numbered $1, 2, \ldots, 7$ going around the heptagon.

Now we make a non-standard definition. We say that the *distance* between two vertices is the length of the shortest path between them on the boundary of the heptagon. Thus the distance between 3 and 6 is three, but the distance between 1 and 6 is only two. All distances between distinct vertices of the K_7 are either one, two or three.

The colouring we used can now be succinctly described as follows. An edge is red precisely when the distance between its end points is one.

This makes it clear that there is no red triangle, but we will take the time to prove (again) that there is no blue K_4. We consider a quadrilateral whose vertices belong to the original heptagon, and call the distances between consecutive vertices a, b, c and d. If the quadrilateral contains the centre of the heptagon then $a + b + c + d = 7$ as shown in figure 6.7. If it does not, then three of the distances sum to the final distance, so we have an equation of the form $a = b + c + d$ as in figure 6.8.

In the first case it is clear that at least one of the distances is 1, and in the second case the fact that $a \leq 3$ shows that $b = c = d = 1$. Either way, the quadrilateral contains at least one red edge of the K_7.

We now know that $8 \leq R(3, 4) \leq 10$. This situation is fairly common in Ramsey theory. A number can easily be trapped between an upper and lower bound, but determining its exact value can be tricky. In exercise 6b we will be able to determine whether $R(3, 4)$ is equal to 8, 9, or 10, but in general we will not always be so lucky.

Now we are in a position to attack $R(4, 4)$.

Problem 6.7

Find an upper bound for $R(4,4)$.

We claim that $R(4,4) \leq 20$. The pattern of the proof is identical to that used in problem 6.5. Suppose the edges of K_{20} are coloured red or blue and choose a vertex v. This vertex has nineteen neighbours so, by the pigeonhole principle, it is contained in at least ten edges that are the same colour. If v has ten red-neighbours, we use the fact that $R(3,4) \leq 10$. The red-neighbours either define a blue K_4 or a red K_3. In the first case we are done, and in the second we add v to the red K_3 to form a red K_4. If, on the other hand, v has ten blue-neighbours, we argue exactly as before but use the fact that $R(4,3) \leq 10$.

In fact, we have just proved that $R(4,4) \leq 2R(3,4)$ so if we manage to push our upper bound for $R(3,4)$ any lower, the upper bound for $R(4,4)$ will automatically come down with it.

Finally we are ready for a truly general result. We know that sufficiently large parties guarantees a group of four mutual friends or strangers, and that a party with twenty guests will certainly suffice. Will a large enough party always contain a group of five mutual friends or strangers, or a group of six or ten or a hundred?

It is not at all obvious that the answer to this question is 'Yes'. Asking for a large monochromatic subgraph is asking for rather a lot of order within the disorder. A monochromatic K_4 only needs six edges to be the same colour, while a monochromatic K_{100} needs $\binom{100}{2} = 4950$. The next two problems will prove that there is an n which guarantees a monochromatic K_{100}. In fact, they will provide us with an (admittedly large) upper bound for $R(100,100)$ and all other Ramsey numbers. The solution to problem 6.8 is from a paper by the Hungarian mathematicians George Szekeres and Paul Erdős ([1]). Both made significant contributions to many areas of combinatorics, and Erdős in particular was hugely influential in the early days of Ramsey theory.

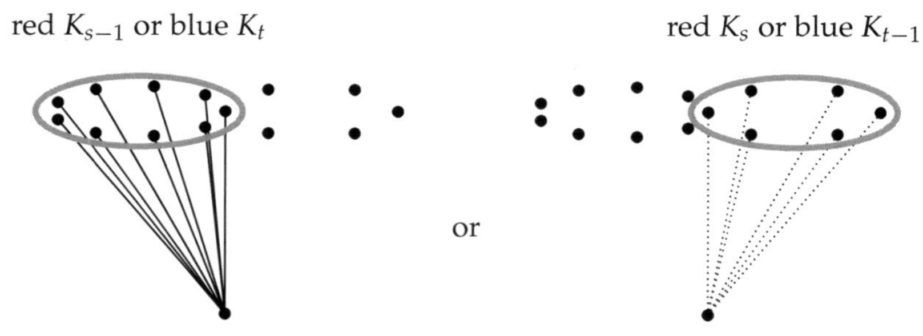

Figure 6.9

Problem 6.8

Prove that $R(s,t) \leq R(s-1,t) + R(s,t-1)$ for all $s, t \geq 3$.

We have seen the argument a number of times already.

Let $N = R(s-1,t) + R(s,t-1)$ and suppose that the edges of K_N are coloured red or blue. We choose any vertex v. This vertex has $N-1$ neighbours, and $N-1 > (R(s-1,t)-1) + (R(s,t-1)-1)$. Therefore, v either has $R(s-1,t)$ red-neighbours or $R(s,t-1)$ blue-neighbours.

The first case is shown on the left of figure 6.9. The red-neighbours of v either contain a blue K_t already, or they contain a red K_{s-1} which can be combined with v to form a red K_s.

The other case is shown on the right of figure 6.9. The blue-neighbours of v either contain a red K_s already, or they contain a blue K_{t-1} which can be combined with v to form a blue K_t.

We need the condition that s and t are at least three, because the proof refers to a red K_{s-1} and blue K_{t-1}, and a K_1 has no edges so is neither red nor blue.

This result, together with the fact that $R(s,2) = s$, allows us to find upper bounds for all Ramsey numbers. We record these upper bounds in the triangular array shown in table 6.1.

We have proved the following result, which we will call the *first form of Ramsey's theorem*:

$R(2,2) = 2$				
$R(2,3) = 3$	$R(3,2) = 3$			
$R(2,4) = 4$	$R(3,3) = 6$	$R(4,2) = 4$		
$R(2,5) = 5$	$R(3,4) \leq 10$	$R(4,3) \leq 10$	$R(5,2) = 5$	
$R(2,6) = 6$	$R(3,5) \leq 15$	$R(4,4) \leq 20$	$R(5,3) \leq 15$	$R(6,2) = 6$

Table 6.1

Ramsey's theorem

For any pair of positive integers (s, t) there exists an integer n such that whenever the edges of K_n are coloured red and blue there is either a red K_s or a blue K_t.

In table 6.1 the Ramsey numbers on each row have a fixed value of $s + t$. If we only include the bounds themselves, and add an extra row, we obtain the triangle shown in table 6.2.

```
              2
           3     3
        4     6     4
     5    10    10     5
  6    15    20    15     6
7   21    35    35    21     7
```

Table 6.2

This triangle looks extremely familiar. It seems to be Pascal's triangle with the outer layer of 1s removed, but we should check that this is not simply a coincidence.

Problem 6.9

Prove that $R(s,t) \leq \binom{s+t-2}{s-1}$ for all $s, t \geq 2$.

We will prove this result by induction. However, we will not induct on s or t. Instead, in order to capture the idea of working down the triangle row by row, we will use induction on $s + t$.

Base

The only way $s+t=4$ with $s,t \geq 2$ is if $s=t=2$. Here we note that $R(2,2)=2$ and that $2=\binom{2+2-2}{2-1}$ as required.

Step

Consider the Ramsey number $R(s,t)$ where $s+t>4$.

If $s=2$ we have $R(s,t)=t=\binom{s+t-2}{s-1}$ as required. The case $t=2$ is identical.

If $s,t \geq 3$ we may use the result from problem 6.8 which said that $R(s,t) \leq R(s-1,t)+R(s,t-1)$. By induction we may assume that both the Ramsey numbers on the right hand side of this inequality are bounded above by the corresponding binomial coefficients.

Thus $R(s,t) \leq \binom{s-1+t-2}{s-2} + \binom{s+t-1-2}{s-1}$.

Now we recall that $\binom{n}{r} = \binom{n-1}{r-1} + \binom{n-1}{r}$ (Pascal's identity) and conclude that $R(s,t) \leq \binom{s+t-2}{s-1}$ as required.

The proof is relatively straightforward, but it is written in a slightly less structured way than we are used to. The induction arguments we have met so far have almost always used the variables n and k. The first is used to refer to the general statement we want to prove, and the second to refer to the specific instance of that statement which we assume holds in order to prove the next one. The proof above can be easily adapted to fit this model. We would phrase the inductive hypothesis as follows:

'If $s+t=n$ then $R(s,t) \leq \binom{s+t-2}{s-1}$'.

Now assume this result holds for some $n=k$, consider a Ramsey number $R(s,t)$ with $s+t=k+1$, and proceed as before.

This approach makes the structure of the argument very clear. However, it has the disadvantage that it introduces the two new variables n and k into a problem which already has plenty of variables.

Our final task in this section will be to prove a lower bound for another Ramsey number. Checking that the construction works is fairly delicate, and it will be good preparation for the exercise which follows it.

Problem 6.10

Prove that $R(3,6) \geq 17$.

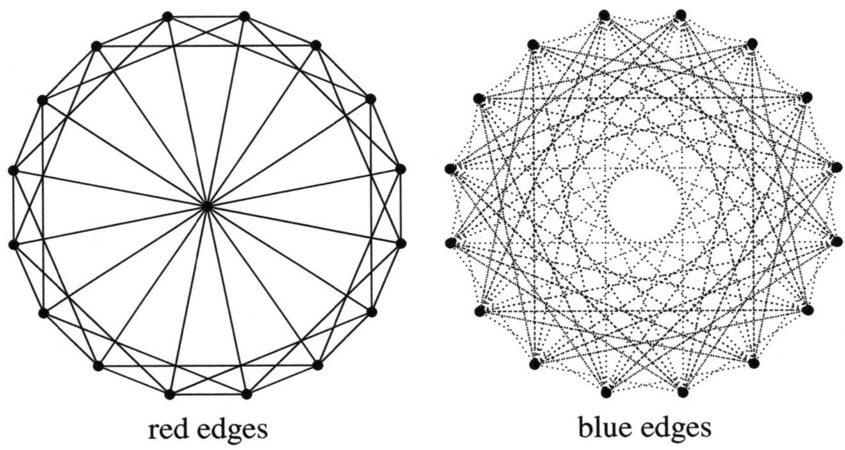

Figure 6.10

We must provide a red-blue colouring of the edges of K_{16} which has no red K_3 or blue K_6.

We will use the vertices of a regular 16-gon with side length 1 as the vertices of our graph. Using the same definition of distance as we did in problem 6.6, we see that the distance between any two distinct vertices is a positive integer less than or equal to 8.

We colour an edge red if the distance between its ends is 1, 3 or 8 and blue otherwise. This gives rise to the colouring shown in figure 6.10. We will discuss this choice once we have checked that it works.

The first task is to check that this graph contains no red K_3. This is not quite as obvious as it was in the K_7 colouring.

Let us suppose that the distances between the vertices of some triangle in K_{16} are a, b and c. If the triangle contains the centre of the 16-gon, then $a + b + c = 16$ as shown in figure 6.11. If it does not, then we have an equation of the form $a = b + c$ as shown in figure 6.12. If an edge of the triangle passes through the centre of the 16-gon both equations hold, since we may assume that $a = 8$.

Now it is easy to check that our colouring does not contain a red triangle, since the equations $a + b + c = 16$ and $a = b + c$ cannot be solved using only 1, 3 and 8.

Checking that this colouring does not contain a blue K_6 is a little more subtle.

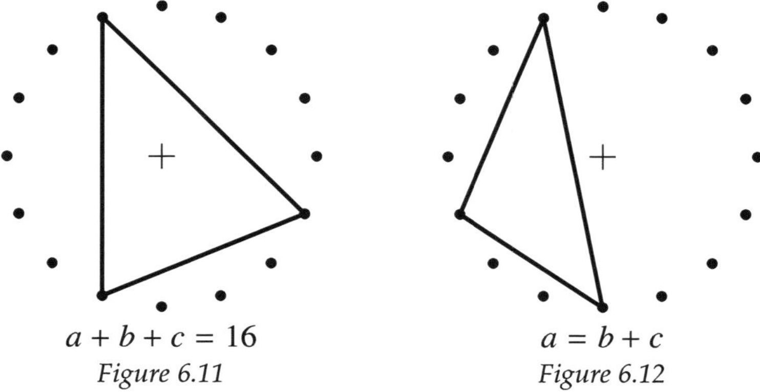

Figure 6.11 *Figure 6.12*

We consider six points from the 16-gon and let the distances between them going clockwise be a, b, c, d, e and f. As before, we either have $a = b + c + d + e + f$ or $a + b + c + d + e + f = 16$ depending on whether or not the hexagon which forms the boundary of the K_6 contains the centre of the 16-gon. This gives two cases.

Case 1

If $a = b + c + d + e + f$, then, since $a \leq 8$ by the definition of distance, we have that one of b, c, d, e, f equals one. So any such hexagon includes a red edge.

Case 2

Suppose that $a + b + c + d + e + f = 16$ and that all the distances on the left belong to the set $\{2, 4, 5, 6, 7\}$ of blue distances. We may not have three or more of these distances greater than or equal to four, so either four or five of them must be equal to two. This gives two subcases.

Case 2a

If the distances are $2, 2, 2, 2, 2$ and 6, then there is only one shape of hexagon to consider. Some of the diagonals of this hexagon have length 8, so the hexagon is not a blue K_6. The hexagon is shown in figure 6.13. The dashed lines are the diagonals of length 8.

Case 2b

If the distances are $2, 2, 2, 2, 4$ and 4, then there are three distinct hexagons to consider. This is because the distances going clockwise

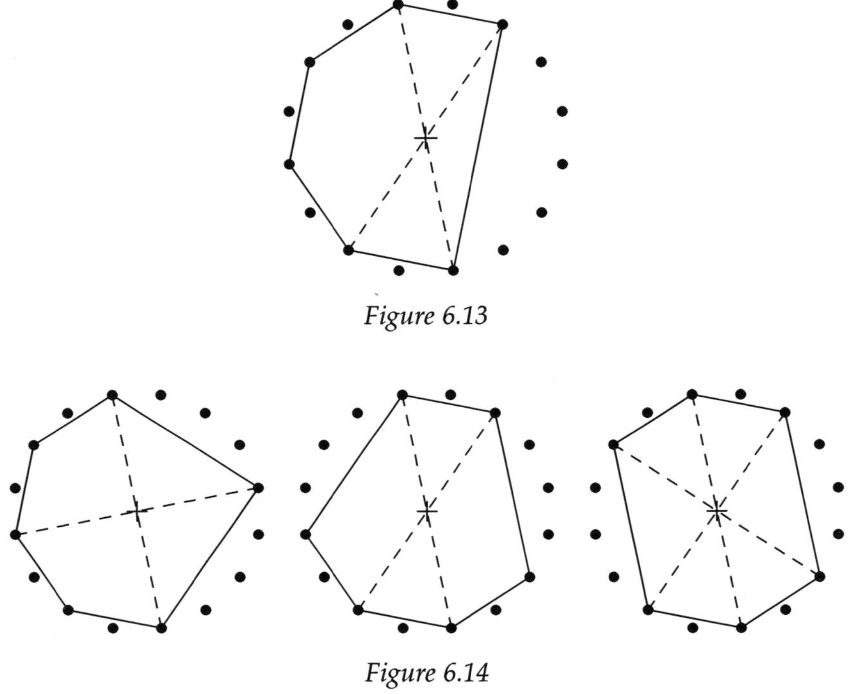

Figure 6.13

Figure 6.14

and starting with a 4 might be 4, 4, 2, 2, 2, 2 or 4, 2, 4, 2, 2, 2 or 4, 2, 2, 4, 2, 2. The possible hexagons are shown in figure 6.14. Each of these hexagons also has diagonals of length 8, so the proof is complete.

The reader might object that colouring an edge red if the distance between its ends is 1, 3 or 8 looks suspiciously like divine inspiration. However, the idea of colouring based only on distance is appealing because it makes the colouring highly symmetric which, in turn, makes it easier to analyse. It is not immediately clear which distances to assign the colour red to, but it is generally easy to tell when a choice will not work. For example, we cannot have lengths 1 and 2 both being red, since this would yield red triangles. Working through possible colouring schemes systematically yields a solution to the problem fairly quickly, as interested readers are welcome to check.

Exercise 6b

1. Prove that $R(3,4) \geq 9$.

2. Look back at the proof that $R(3,4) \leq 10$. Now suppose that there is a red-blue colouring of the edges of K_9 which contains no red K_3 or blue K_4.

   (a) How many red-neighbours must each vertex in this coloured K_9 have?

   (b) How many red edges must there be in the graph?

   (c) What does this prove?

3. Prove that $R(s,s) \leq 4^{s-1}$.

4. Prove that $R(3,5) \geq 14$.

5. Prove that $R(3,6) \leq 19$. (*Hint: this is similar to question* 2)

6. Prove that $R(s,s) > (s-1)^2$.

7. Prove that $R(4,4) \geq 18$.

6.3 $R(5,5)$ and $R(6,6)$

In question 2 of exercise 6b you were asked to prove that $R(3,4) \leq 9$. If we combine this with the fact that $R(s,t) \leq R(s-1,t) + R(s,t-1)$, we obtain the slightly better upper bounds for Ramsey numbers shown in table 6.3.

This allows us to conclude that $R(5,5) \leq 64$. Working the other way, question 6 of exercise 6b shows that $R(5,5) > 16$, which leaves 48 possible numbers which $R(5,5)$ could be. It is not immediately obvious how to improve either of these bounds. However, reducing the problem of finding $R(5,5)$ to a finite search seems like a big step forward. It may not be particularly elegant, but surely at this point a computer could

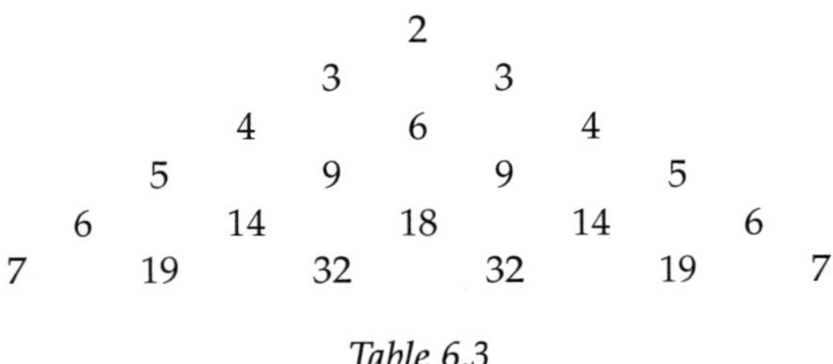

					2					
				3		3				
			4		6		4			
		5		9		9		5		
	6		14		18		14		6	
7		19		32		32		19		7

Table 6.3

work through these cases one by one, just as was done in the proof of the four-colour theorem.

Let us consider what this task would entail. If we wanted to check that $R(5,5) \leq 63$, we would need to list all possible red-blue edge colourings of K_{63} and then, for each of these colourings, check that at least one of the K_5 subgraphs was monochromatic. How hard can it be?

The short answer is 'very'. For each colouring of K_{63}, a naïve approach would involve checking $\binom{63}{5} = 7\,028\,847$ possible K_5 graphs. However, the real problem arises when we count the number of possible colourings of K_{63}. There are $\binom{63}{2} = 1953$ edges, each of which might be one of two colours. This gives 2^{1953} possible colourings. Multiplying these together we have approximately 5.73×10^{594} cases to check.

This number is completely out of reach, even for a supercomputer. At the time of writing, a hundred billion state-of-the-art supercomputers working for a hundred billion years would come nowhere near to checking even one percent of the cases: a brute force computational approach is unlikely to be of much use.

Using computers to attack the upper bound on Ramsey numbers is extremely hard, but they can be helpful in improving lower bounds. To show that $R(5,5) \leq 63$ we must check that every colouring of K_{63} has a monochromatic K_5, while showing that $K(5,5) > 41$ only requires us to find one colouring of K_{41} which does not have a monochromatic K_5. If we restrict our attention to colourings which are only based on distance, then the computation becomes much more manageable since the kind of analysis we used in problem 6.10 is fairly easy to automate. For example, in 1976 Hanson used a computer to show that $R(5,5)$ is indeed greater than 41. In K_{41} the distances between vertices are integers less than or equal to 20, and if we colour an edge red if and only if the distance between its ends is $1, 3, 5, 8, 12, 13, 16, 17, 18$ or 19, then the resulting graph does not contain a monochromatic K_5 as required.

Further progress has been made using theoretical methods, rather than computational, but the exact value of $R(5, 5)$ is still unknown. If you ask the question:

'What is the smallest number of people you need at a party to guarantee the existence of a group of five friends or five strangers?' then the best answer is currently: 'Nobody knows exactly: somewhere between 43 and 49 inclusive.'

The situation for $R(6, 6)$ is even more unsatisfactory. All that is known is that $102 \leq R(6, 6) \leq 165$. There is a famous story about Paul Erdős, which serves to illustrate how hard improving these bounds is. When asked about finding exact values of Ramsey numbers Erdős replied:

'Suppose aliens invade the earth and threaten to obliterate it in a year's time unless human beings can find the Ramsey number for red five and blue five. We could marshal the world's best minds and fastest computers, and within a year we could probably calculate the value. If the aliens demanded the Ramsey number for red six and blue six, however, we would have no choice but to launch a pre-emptive attack.'

6.4 A better lower bound for $R(s, s)$

While knowing the values of numbers like $R(5, 5)$ and $R(6, 6)$ is certainly of interest, it is perhaps even more interesting to ask how the number $R(s, s)$ behaves for larger values of s.

Currently all we know is that $(s-1)^2 < R(s, s) < \binom{2s-2}{s-1}$. The lower bound is a quadratic polynomial, and it it is not hard to show that the upper bound approaches $k \times 4^s$ for some constant k. As s gets larger, the gap between these bounds becomes more and more embarrassing. Do the actual Ramsey numbers behave like some polynomial function (meaning they grow fairly slowly), or do they behave like an exponential function (meaning they grow very fast)?

In 1947 Erdős proved that $\sqrt{2}^s < R(s, s)$. This is an exponential lower bound for $R(s, s)$ so it shows that the growth rate of Ramsey numbers is more or less exponential. This is a huge leap forward, though, as we shall see, it still leaves some important questions unanswered.

We will approach Erdős' result in two stages. First we will prove a rather strange looking lower bound for $R(s, s)$, which is where the important combinatorics happens. Then we will use some algebra to convert the bound into its memorable exponential form.

Problem 6.11

Prove that if $2\binom{n}{s} < 2^{\frac{s(s-1)}{2}}$, then $R(s,s) > n$.

The crucial idea in Erdős' lower bound is to fix two numbers n and s and then perform two calculations:

- find the number of 2-colourings of the edges of K_n; call this X.
- find the number of 2-colourings of the edges of K_n which contain a monochromatic K_s; call this Y.

If $X > Y$, then there must be at least one red-blue colouring of K_n with no monochromatic K_s, so $R(s,s) > n$.

It turns out that it is tricky to find Y exactly, but it is possible to find an overestimate for it. If this overestimate is still less than X, then the conclusion follows as before.

Finding X is straightforward. K_n has $\frac{n(n-1)}{2}$ edges, each of which can be one of two colours so there are $2^{\frac{n(n-1)}{2}}$ colourings.

To calculate an overestimate for Y, we first choose s of the vertices from the K_n, which can be done in $\binom{n}{s}$ ways. These vertices form a K_s, and we colour all its edges the same colour. There are two colours to choose from so we have $2 \times \binom{n}{s}$ ways to create a monochromatic K_s. Now we colour the remaining edges in the K_n. There are $\frac{n(n-1)}{2} - \frac{s(s-1)}{2}$ of these, each of which can be coloured in one of two ways. Hence $Y < 2\binom{n}{s}2^{\frac{n(n-1)-s(s-1)}{2}}$.

This is an overestimate since we have counted colourings of K_n which contain more than one monochromatic K_s more than once. Nevertheless, we see that $R(s,s) > n$ provided $2^{\frac{n(n-1)}{2}} > \binom{n}{s}2^{\frac{2+n(n-1)-s(s-1)}{2}}$ and this, after a little manipulation, gives $2^{\frac{s(s-1)}{2}} > 2\binom{n}{s}$.

This result provides lower bounds for $R(s,s)$ but provides no efficient way of constructing colourings of K_n which do not contain monochromatic copies of K_s. The proof shows that such colourings exist without exhibiting any of them.

Now we will convert the bound in problem 6.11 into something more manageable.

Problem 6.12

Prove that $2\binom{n}{s} < 2^{\frac{s(s-1)}{2}}$ and hence that $R(s,s) > n$, provided $n \leq \sqrt{2}^s$.

We need to prove that

$$2 \times \frac{n(n-1)\ldots(n-s+1)}{s!} < 2^{\frac{s^2}{2}} \times 2^{-\frac{s}{2}}.$$

We may multiply by $\frac{s!}{2}$ and then overestimate the left hand side, so it suffices to show that

$$n^s < 2^{\frac{s^2}{2}} \times (s! \times 2^{-\frac{s}{2}} \times \frac{1}{2}).$$

Since $s! > 2^s$, the bracket on the right hand side is greater than 1. Therefore, it suffices to prove that $n^s \leq 2^{\frac{s^2}{2}}$.

Now we use the fact that $n \leq 2^{\frac{s}{2}}$, and the result follows.

This result means that in 1947 it was known that $\sqrt{2}^s < R(s,s) < 4^s$. The next challenge for the mathematical community was to try to narrow the gap between these exponential bounds. In particular, the question of whether the 4 in the upper bound can be reduced, or the $\sqrt{2}$ in the lower bound can be increased, has been extensively studied but significant progress is still awaited.

6.5 More colours

If you managed to answer the last question in exercise 6a, then the idea of colouring the edges of a complete graph with more than just two colours is already familiar.

We will use $R(s,t,u)$ to denote the least n such that whenever the edges of K_n are coloured red, blue or green, there exists a red K_s, a blue K_t or a green K_u. The question in exercise 6a showed that $R(3,3,3) \leq 17$.

Problem 6.13

Prove that $R(4,3,3) \leq R(4,6)$.

We know that the number $R(4,6)$ exists. Indeed, while we do not know its exact value, we know it is at most $\binom{4+6-2}{4-1} = 56$. The idea behind the proof is to suppose that the edges of $K_{R(4,6)}$ are coloured red, blue and green but that we are wearing turquoise spectacles. The effect of these spectacles is to make blue and green edges indistinguishable: they all look turquoise. The graph is large enough to guarantee that we either see a red K_4 or a turquoise K_6. In the first case we are done. In the second we restrict our attention to the K_6 and remove our spectacles. We now see a copy of K_6 where all the edges are blue or green. Since $R(3,3) = 6$, we are sure to have a blue or a green triangle, so the problem is solved.

Exercise 6c

1. Prove that $R(s,t,u)$ exists for all $s,t,u \geq 2$.

2. Prove that $R(s,s,s) \leq 4^{s^2}$.

3. Prove that $R(s,s,s) \geq \sqrt{3}^s$.

4. Let $R(s,t,u,v)$ be the least n such that when the edges of K_n are coloured red, blue, green or yellow there is a red K_s, a blue K_t, a green K_u or a yellow K_v. Prove that $R(s,t,u,v)$ exists for all $s,t,u,v \geq 2$. Can you generalise?

5. Prove that $R(s,s,s,s) \leq 4^{4^s}$.

6.6 To infinity...

We know that if we 2-colour the edges of a sufficiently large complete graph, then we can always find a large monochromatic subgraph. Now we will move on to consider the analogous problem for infinite graphs and ask whether every 2-coloured infinite complete graph contains an infinite monochromatic subgraph.

More precisely, suppose we have a complete graph whose vertices are the natural numbers $1, 2, 3, \ldots$. An edge in this graph is a pair of integers $\{x, y\}$ and a colouring is a function that assigns the colours red or blue to these pairs. To keep things concrete we consider two specific examples:

Colouring A: $\{x, y\}$ is red if xy has an odd number of factors and blue otherwise.

Colouring B: $\{x, y\}$ is red if $x + y$ has an odd number of factors and blue otherwise.

For colouring A it is easy to find infinite monochromatic sets. If we consider the set of prime numbers, then the product of every pair has exactly four factors, so every edge is blue. On the other hand, if we consider the set of squares, then the product of every pair is again a square number, so every edge is red.

Finding large monochromatic sets for colouring B is much more troublesome and feels like a hard number theory problem, though, as we shall see, the number theory turns out to be entirely irrelevant.

We need to introduce some more notation. Since we called the complete graph on n vertices K_n, it might seem sensible to call an infinite complete graph K_∞. However, since the vertices in the previous example are numbered and the colouring depends on the numbering, the graph has more structure than a simple (unlabelled) K_∞. Instead we will use the notation $X^{(2)}$ to denote the set of (unordered) pairs of members of X. Thus, $X^{(2)}$ is the set of edges of the complete graph with vertex set X. The examples above concern 2-colourings of $\mathbb{N}^{(2)}$, and our question is whether every 2-colouring of $\mathbb{N}^{(2)}$ has an infinite monochromatic set.

The first or *finite* form of Ramsey's theorem on page 206 certainly ensures that every 2-colouring of $\mathbb{N}^{(2)}$ has arbitrarily large monochromatic sets. Indeed, if X is any infinite set, then we can take a finite subset with more than $R(s, s)$ members and find a monochromatic set of size s in that subset. However, the existence of arbitrarily large structures does not imply the existence of infinite structures with the same properties. For example, the integers contain arbitrarily long sequences of consecutive composite numbers, but none of these sequences is infinite.

This means that we cannot hope to deduce the answer to our question from what we have already proved. We will need a new argument.

Problem 6.14

Prove that every 2-colouring of $\mathbb{N}^{(2)}$ has an infinite monochromatic set.

This is the infinite version of the statement '$R(s,s)$ exists for all s', so we look to the proof of that result for inspiration.

The key steps in the finite case went something like this:

- choose an arbitrary vertex v in a large complete graph.
- by the pigeonhole principle, there is a large set S of points joined to v by edges of the same colour.
- if S is large enough (of size at least $R(s-1,s)$) then, by induction, good things happen inside S.

We would like to adapt this proof to our new, infinite, setting. However, we cannot use normal induction to get all the way to infinity. Fortunately, in the argument above the phrase 'by induction' was really just shorthand for 'repeat the same argument a large number of times' so the finite proof can be summarised:

- choose an arbitrary vertex v in a large complete graph.
- by the pigeonhole principle, there is a large special set of points joined to v by edges of the same colour.
- repeat this process inside the special set a large number of times.

The plan now is simply to replace every instance of the word 'large' with the word 'infinite'.

Suppose every edge in $\mathbb{N}^{(2)}$ is coloured red or blue.

Pick an integer v_1, for example, $v_1 = 1$.

By the pigeonhole principle, there is an infinite set, S_1, of integers such that all the edges joining v_1 to members of S_1 are the same colour. This colour is either red or blue, but, since we do not know which, we will call it colour c_1.

From now on we will only work within the infinite set S_1.

We choose an integer $v_2 \in S_1$.

As before, there is an infinite set, $S_2 \subset S_1$ such that all the edges joining v_2 to members of S_2 are the same colour. We call this colour c_2.

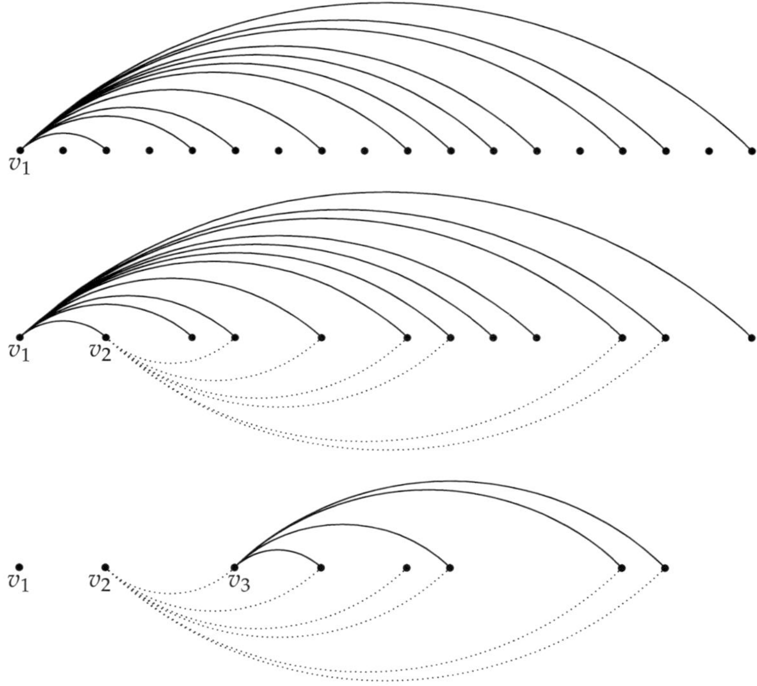

Figure 6.15

We note that every member of S_2 is joined to v_1 by an edge of colour c_1, as well as being joined to v_2 by an edge of colour c_2.

Now we choose $v_3 \in S_2$ and find an infinite set $S_3 \subset S_2$ such that all the edges from v_3 to members of S_3 have colour c_3. The first three stages of the process are shown in figure 6.15. The colours c_1, c_2 and c_3 are represented by solid, dotted and solid lines respectively.

We now repeat this process choosing $v_i \in S_{i-1}$ and $S_i \subset S_{i-1}$ such that edges from v_i to members of S_i have colour c_i. Since all the S_i are infinite, we never run out of integers. Thus, at the end of all time, we obtain three infinite sequences:

(i) a sequence $v_1, v_2, v_3, \ldots$ of integers;

(ii) a sequence $S_1 \supset S_2 \supset S_3 \supset \ldots$ of nested infinite sets;

(iii) a sequence $c_1, c_2, c_3, \ldots$ of colours each of which is either red or blue.

Our construction ensures that, for any n, the edge (v_1, v_n) has colour c_1, the edge (v_2, v_n) has colour c_2 and in general (v_i, v_n) has colour c_i for all $i < n$.

Now we focus on the list of colours. At least one of the colours occurs infinitely often, and we call this colour C. We now form an infinite set, S, of integers by including v_i in S if $c_i = C$.

By construction, every edge joining two members of S has colour C. Therefore, S is an infinite monochromatic set as required.

An important feature of the proof is that it gives no indication as to how one might construct an infinite monochromatic set. If we look back at the two example colourings of $\mathbb{N}^{(2)}$ on page 217, we now know that colouring B does give rise to infinite monochromatic sets, but we certainly cannot exhibit one. In fact, no explicit construction of a monochromatic set is known for this colouring.

6.7 ...and back

In our solution to problem 6.14 we noted that, since $R(s,s)$ exists for all s, any 2-colouring of $\mathbb{N}^{(2)}$ contains arbitrarily large monochromatic sets. However, we took care to stress that the existence of large monochromatic sets does not imply the existence of infinite ones. In other words, we cannot use the finite form of Ramsey's theorem to prove the infinite form. In fact, our proof of the infinite form made no reference to the finite form of the result at all, so it would not have been ridiculous to discuss the infinite form of the theorem first.

It may, therefore, come as a surprise that the infinite form can be used to prove the finite form. The infinite form of the theorem is therefore a strictly stronger result. Of course, we already have a valid proof of the finite form, so in a sense there is no need for another. However, having more than one proof of a theorem is often enlightening and deducing the finite form directly from the infinite form will prepare us for a further generalisation.

Our strategy will be to assume, for contradiction, that the finite form is false. We will then use this assumption to build a counterexample to the infinite form of the theorem. Since we know the infinite form of the theorem is true, we may conclude that the finite form cannot be false.

We will use $[N]$ to denote the set $\{1, 2, 3, \ldots, N\}$, and, as before, we will use $[N]^{(2)}$ to denote the set of pairs $\{a, b\}$ where a and b are in $[N]$.

The set $[N]^{(2)}$ is just the edge set of a complete graph K_N where the vertices have been labelled with the numbers 1 to N. This means we can rephrase the claim that $R(s,s)$ always exists as follows.

For all s there exists an N such that every 2-colouring of $[N]^{(2)}$ contains a monochromatic copy of K_s.

This is a useful rephrasing since we now want to assume that this statement is false and derive a contradiction. In particular, we assume the following.

There exists a number s such that, for all N, there is a 2-colouring of $[N]^{(2)}$ with no monochromatic K_s. $(\star)$

We will call colourings with a monochromatic K_s *good*, and other colourings *bad*. In these terms the existence of $R(s,s)$ means that, for any given s, all sufficiently large colourings are good, while $(\star)$ claims that, for some values of s, there are arbitrarily large bad colourings.

Now we will use these bad colourings to construct a colouring on the whole of $\mathbb{N}^{(2)}$ which contains no monochromatic K_s. Since every colouring of $\mathbb{N}^{(2)}$ contains an infinite monochromatic set, this leads to a (dramatic) contradiction.

We let C_i^1 be a bad colouring of $[i]^{(2)}$. By $(\star)$, such a colouring exists for all i, so we obtain an infinite sequence $C_1^1, C_2^1, C_3^1, \ldots$ of bad colourings on larger and larger sets. (We will see the reason for the superscripts later.)

We would like to be able to stitch all these colourings together to form a colouring on $\mathbb{N}^{(2)}$. Unfortunately, the different C_i^1 may assign different colours to the same edge so we need to tread carefully.

We now list the edges of $\mathbb{N}^{(2)}$ in some order $e_1, e_2, e_3 \ldots$.

For example we might use the ordering:

$$\{1,2\}, \{1,3\}, \{2,3\}, \{1,4\}, \{2,4\}, \{3,4\}, \{1,5\} \ldots.$$

Here (a,b) comes before (c,d) if $b < d$ and (a,b) comes before (c,b) if $a < c$.

It is clear that the list $e_1, e_2, \ldots$ contains every edge of $\mathbb{N}^{(2)}$ at some point, so we may assign colours to these edges one at a time to form a single infinite colouring which we will call C.

First we look at the colours assigned to e_1 by C_1^1, C_2^1, C_3^1, and so on. By the pigeonhole principle, at least one colour, which we call c_1, occurs infinitely often.

We let C assign this colour to e_1. Now we look at the sequence of colourings $C_1^1, C_2^1, C_3^1, \ldots$ and remove any colourings which did not assign colour c_1 to edge e_1. This still leaves an infinite sequence of colourings which we call $C_1^2, C_2^2, C_3^2, \ldots$.

Among the C_i^2 there are infinitely many colourings which assign e_2 the same colour. We call this colour c_2 and let C assign c_2 to e_2. Now we remove all colourings which do not assign c_2 to e_2 from the list of bad colourings. This still leaves an infinite sequence of colourings $C_1^3, C_2^3, C_3^3, \ldots$.

Each of the colourings C_i^3 assigns c_1 to e_1 and c_2 to e_2 by construction. We can also find infinitely many which assign some colour c_3 to e_3. We let C assign c_3 to e_3 and remove colourings which do not assign e_3 this colour from our list. We repeat this process for each edge in turn.

Every edge in the list $e_1, e_2, \ldots$ is eventually assigned a colour, so we have constructed a colouring of the whole of $\mathbb{N}^{(2)}$.

We claim that this colouring does not contain a monochromatic K_s.

To prove the claim, we suppose that C contains a monochromatic K_s and that the last edge in this K_s is e_M for some M, where last means latest in the list $e_1, e_2, \ldots$ of the edges.

We consider the colouring C_1^{M+1}.

By construction, this colouring assigns colour c_i to edge e_i for all $i \leq M$. In particular it contains the same monochromatic K_s as C does. However, all the C_i^j are bad colourings, so contain no monochromatic K_s. This contradiction proves that C contains no monochromatic K_s. This, in turn, contradicts the infinite form of Ramsey's theorem, so the claim $(\star)$ must have been false.

We have proved for the second time that $R(s,s)$ exists for all s. The proof is fairly subtle, and you are advised to study it carefully before attempting the following exercise. It is also worth noting that we never actually used the fact that the colourings we were using were 2-colourings, merely that the number of colours was finite at each stage. Therefore, the proof in fact shows that Ramsey numbers exist for any finite number of colours.

Exercise 6d

1. Suppose that the set $[N]^{(2)}$ is k-coloured. We say a set of $S \subset [N]$ is monochromatic if all the edges joining pairs of integers in S are

the same colour. A monochromatic subset of $[N]$ is called a *nice* set if the number of elements in the set is greater than or equal to the smallest member of the set. More formally, the monochromatic set $S = \{v_1, v_2, \ldots, v_n\}$ satisfying $v_1 < v_2 < \cdots < v_n \leq N$ is nice if $n \geq v_1$.
Prove that for any k there exists an N such that every k-colouring of $[N]^{(2)}$ contains a nice set. This is known as the *strengthened Ramsey theorem*.

6.8 Proving the unprovable

At the start of this chapter, we noted that Ramsey proved his famous theorem in the early pages of a paper about mathematical logic. In particular, Ramsey was interested in finding ways to decide whether or not a given statement is *provable*.

The word 'provable' has a slightly odd ring to it. Most mathematicians would say they spend their time trying to find out whether or not things are *true*, and generally consider something to be true once a proof of it has been found.

However, it turns out that the notions of truth and provability are not the same. In the discussion that follows, we hope to give some indication of the differences between them, though any attempt to compress so many ideas into so little space can only be partially successful.

We begin by asking: 'What do we mean by a proof in mathematics?' and leave the question of 'What do we mean by a true statement?' until later.

A reasonable first attempt at a definition might be something like:

'A sequence of statements leading to a desired conclusion, where each one follows logically from previously established statements.'

However, this raises some important questions. It suggests that if we look at a proof, which we assume for simplicity has exactly one statement per line, then we should be able to ask 'Why are we allowed to write that?' of every line in the proof.

Often a line in a proof will be a consequence of one or more earlier lines in the proof, but to make sense of this we first need to agree what it means for one statement to follow logically from another. In other words, we need to agree on the rules of logic. Fortunately there is wide agreement as to what the 'correct' rules of logic are. They include statements like 'If we have already written not(not A), then we may write A', 'If we have already written A, then we may also write (A or B)' and 'If we have already written (A implies B) and A, then we may write B'. These logical rules can easily be listed on a single A4 page.

However, the rules of logic alone do not allow us to do very much mathematics. For example, if we are studying geometry, we will want to refer to things called lines, points, circles and so on. The rules of logic make no mention of these objects, so we will need to agree that some statements about these objects are true, without requiring further justification. Such foundational assertions are called *axioms* and which axioms mathematicians use depends on what they are studying. An example of a geometrical axiom is the statement that it is possible to draw exactly one straight line through two distinct points. The standard collection of geometrical axioms (which can also be listed on a single page) form what is called the *theory of plane geometry*. If we are examining a geometrical proof and ask 'Why are we allowed to write this?' of a line which happens to be an axiom, then the answer is 'We simply all agree that we can.'

People who think that mathematics is a source of immutable truths may find this unsettling, but there is a sense in which every statement in mathematics is of the form: 'If we allow certain basic assumptions, then we may conclude that the following theorem holds.'

Two other theories which will be of interest to us are the theory of arithmetic and the theory of sets.

In the first case the objects under discussion are the natural numbers. Here are the axioms.

- 1 is a natural number.
- If n is a natural number, then so is $n+1$.
- If $n+1=m+1$, then $n=m$.
- There is no natural number n such that $n+1=1$.

As well as these basics, the theory of arithmetic allows the use of induction. In particular, for any proposition P which might or might not be true of a natural number, it is axiomatic that if $P(1)$ is true and $P(k)$ implies $P(k+1)$, then $P(n)$ is true for all $n \in \mathbb{N}$.

These axioms are referred to as the axioms of *Peano arithmetic* or simply *PA*. Many statements in number theory, like the fact that every natural number has a unique prime factorisation, or the fact that the sum of two odd squares is never a square, can be proved in PA.

At a formal level, the bedrock of the theory of numbers is a handful of logical rules and a handful of axioms about addition. The fact that these modest beginnings have such rich and varied consequences is one of the great beauties of mathematics.

It should be stressed that we are not attempting to give an introduction to formal logic, and that many details have deliberately been omitted. Our aim is only to give some indication of the type of thinking involved in this area of mathematics, and how it relates to combinatorics in particular.

For our purposes, a crucial fact is that the finite form of Ramsey's theorem, can, after some effort, be expressed and proved in the language of Peano arithmetic. If a pedantic toddler were to ask 'Why?' about every line of our proof of that theorem, we could eventually reduce it to a lengthy sequence of logical deductions from the axioms listed above.

Few people would ever write out such a sequence of deductions in full, but it is important that it can be done in principle.

PA is a remarkably rich theory. However, it lacks the capacity to talk explicitly about infinite things. Thus the statement 'The set of primes is infinite' must, in Peano arithmetic, be translated as 'For any n there is a prime greater than n'.

Mathematicians often do want to reason about infinite objects, and for that they turn to the *theory of sets*.

While PA has about half a dozen axioms relating to natural numbers, the theory of sets consists of about a dozen axioms relating to sets. These are then combined with the rules of logic to form proofs. The theory of sets is of particular interest because it turns out that almost all of modern mathematics can be rephrased in terms of sets.

For example, while the axioms of geometry talk about lines as objects in their own right, it is also possible to *define* a line as the *set* of pairs of numbers (x, y) which satisfy an equation of the form $ax + by = c$ for some numbers a, b and c. Circles can be similarly defined, and the whole of geometry can be converted into a collection of statements about sets of pairs of numbers.

This does not mean that mathematicians studying geometry no longer *think* of lines and circles as individual objects, but it does mean that, if pushed, mathematicians could convert geometric proofs into sequences of deductions from the axioms of set theory. In practice this exercise would be unbearably tedious and would be considered a waste of time, but it can be done in principle.

Similarly, statements about natural numbers can be translated into statements about sets, as indeed can statements about almost all other objects in mathematics.

This is of theoretical interest since it means that the axioms of set theory could reasonably be considered to be the axioms underlying (nearly) all mathematics. Moreover, whenever we reason about things that are infinite, or things that occur once something has been done infinitely often, we should expect the axioms of set theory to be lurking somewhere deep below the surface.

Uniting all of mathematics under the banner of set theory may be an appealing idea, but it could also be argued that a correct statement about the natural numbers, somehow *ought* to be deducible from the Peano axioms. After all, PA aims to capture how we think the natural numbers should behave.

Fortunately, the Peano axioms can themselves be translated into the language of set theory, so any proof in PA can be translated into a proof in the theory of sets. (The Peano axioms do not correspond to the axioms of set theory, rather, they are *theorems* which can be *deduced* from the axioms of set theory.)

However, there is no analogous way of translating statements in set theory into statements in PA.

This raises an initially unsettling question. Might there be statements about the natural numbers which are true, and which can be proved using set theory, but which cannot be proved using Peano arithmetic? It turns out that the answer is 'Yes'.

To spell out what this means we must return to the question of what we mean by a *true* statement. This is an expansive question and we will not give a full answer. Instead we content ourselves with making two observations about the notion of truth in mathematics. The first is that if we are prepared to say that the axioms in a particular theory are all true, then we must also say that all the theorems which can be proved from those axioms are true (because we agree that that is how the rules of logic work). The second is that when it comes to statements about the natural numbers, we may call a statement true if it has no counterexamples. This definition of truth means that there is an astonishingly wide variety of true statements. In fact, it is not all that surprising that a small collection of axioms might not be able to prove all of them. The fact that PA cannot prove certain true statements only seems surprising because PA happens to be remarkably good at proving results which we consider to be interesting. Once the point that PA might not prove every true statement has been conceded, it is also unsurprising that another collection of axioms might be able to prove some things which PA cannot. The axioms of set theory turn out to be strictly more powerful than PA, so they can prove certain statements which are inaccessible to PA.

In 1931 Kurt Gödel constructed an explicit example of a statement about the natural numbers which is true, but not provable in PA. In the light of the preceding discussion, the remarkable thing here is that Gödel was able, using axioms from outside PA, to prove that his statement was *unprovable* in PA. In fact, Gödel's work goes much further than just cooking up a true statement which does not follow from the Peano axioms, but we will resist the temptation to discuss his results here.

Gödel's true yet unprovable statement seemed a rather terrifying entity at the time, but some mathematicians took comfort in the fact that the statement was hopelessly contrived. Gödel had designed it specifically to ensure it was unprovable, and it seemed plausible that such a weird statement simply would not crop up in the course of 'normal' mathematical research. Sadly they were mistaken.

In 1977 Jeff Paris and Leo Harrington showed that a certain theorem in Ramsey theory was not provable in PA ([5]). Their unprovability result is beyond the scope of this book, but we are in a position to show that this unprovable result is in fact true. Of course, our proof will necessarily use notions which are not expressible in PA, but if we are happy to use notions of infinity, which come from set theory, then this is no obstacle. In fact, diligent readers have already done the work.

If you managed to complete the previous exercise, by using an argument like the one in section 6.7 to prove the strengthened Ramsey theorem, then congratulations: you proved something which has been proved to be unprovable!

Appendix

We have aimed to keep the mathematical prerequisites required for this book to a minimum. Certainly a reader familiar with [4] will be well prepared to tackle this volume. We list some notational conventions below. Other notation used in the text is entirely standard. Words defined in the text are listed in the index.

Notation

$\mathbb{N}$	The set $\{1, 2, 3, \ldots\}$
$[N]$	The set $\{1, 2, 3, \ldots, N\}$
$\lvert X \rvert$	The number of elements in a set X.
$X^{(2)}$	The set of pairs of elements of X.
$\binom{n}{r}$	The number of combinations without replacement of r things from n. This is given by $\dfrac{n!}{r!(n-r)!}$.
$\binom{n}{a_1\ a_2\ \ldots a_k}$	The number of ways of dividing a set of size n into k subsets with sizes $a_1, a_2, \ldots, a_k$ (where the a_i sum to n). This is given by $\dfrac{n!}{a_1!a_2!\ldots a_k!}$.

$\left(\!\binom{n}{r}\!\right)$	The number of combinations with replacement of r things from n. This is given by $\binom{n+r-1}{r} = \dfrac{(n+r-1)!}{(n-1)!r!}$.
C_n	The cyclic group of order n. It consists of the rotational symmetries of a regular n-gon.
D_n	The dihedral group of order $2n$. It consists of the rotational and symmetries of a regular n-gon.
K_n	The complete graph on n vertices.
S_n	The symmetric group on n elements. It consists of all permutations of $\{1, 2, \ldots n\}$.
Ω	The set of raw colourings of a particular configuration.
Ω_π	The set of colourings of a configuration which are invariant under the transformation π.
$x_1^{n_1} x_2^{n_2} \ldots x_r^{n_r}$	The cycle type of a permutation with n_i cycles of length i for each $1 \leq i \leq r$.
Sameness	When we refer to two graphs or groups being *the same* we actually mean that they are *isomorphic* in some appropriate sense. Giving precise definitions in the text seemed unnecessarily formal, these can be found in any undergraduate textbook on graph theory and group theory respectively.

Useful results

Pascal's identity

$$\binom{n}{r} = \binom{n-1}{r-1} + \binom{n-1}{r}$$

The left hand side of this equation is the number of r-element subsets of $\{1, 2, \ldots, n\}$. We divide these into those subsets which contain n and those which do not. If n is in our subset, then we must choose the remaining $r-1$ elements from $\{1, 2, \ldots, n-1\}$, which can be done in $\binom{n-1}{r-1}$ ways, and if n is not included, then we must choose all r elements from $\{1, 2, \ldots, n-1\}$, which can be done in $\binom{n-1}{r}$ ways. These two possibilities cover all the r-element subsets and the result follows.

Sum to infinity of a geometric progression

Consider $S = 1 + x + x^2 + x^3 + \ldots$ and note that $S = 1 + xS$, which can be solved to yield

$$S = \frac{1}{1-x}.$$

For our purposes this result is useful when combined with the concept of a *generating function* (see below).

This argument assumes that the infinite sum S has some definite value. It can be proved that this is the case provided that $-1 < x < 1$, but the details, though not difficult, are omitted. Readers are referred to any introductory analysis textbook.

Composition of functions

Suppose we have four sets W, X, Y, Z and three functions

$$h : W \to X,\ g : X \to Y,\ f : Y \to Z$$

where the notation $h : W \to X$ means that h maps elements of W to elements of X.

We denote the composite function $f(g(x))$ by $f \circ g(x)$ and claim that composition is *associative*, that is, that

$$(f \circ g) \circ h = f \circ (g \circ h).$$

To prove that these two functions are equal we note that they are both from W to Z and that for any $w \in W$ we have the two sequences of equations

$$f \circ (g \circ h)(w) = f(g \circ h(w)) = f(g(h(w)));$$
$$(f \circ g) \circ h(w) = f \circ g(h(w)) = f(g(h(w))).$$

For our purposes the most important consequence of this result is that composition of symmetry transformations is associative. This is clear since a symmetry of a shape can be regarded as a function mapping points in space to other points in space.

Useful techniques

Principle of inclusion and exclusion (PIE)

Suppose we wish to find the number of elements in the union of some sets $A_1, A_2, \ldots A_n$.

For $n = 2$ and 3 we have

$$|A_1 \cup A_2| = |A_1| + |A_2| - |A_1 \cap A_2|;$$

$$|A_1 \cup A_2 \cup A_3| = |A_1| + |A_2| + |A_3| - |A_1 \cap A_2| - |A_2 \cap A_3| - |A_3 \cap A_1| + |A_1 \cap A_2 \cap A_3|.$$

In general (briefly allowing ourselves to speak of 'adding sets' when we mean adding the numbers of elements in those sets) the pattern is

- add the sets one at a time;
- subtract the intersections taken two at a time;
- add the intersections taken three at a time;
- subtract the intersections taken four at a time and so on.

Formally, if we set $A_{ij} = A_i \cap A_j$, $A_{ijk} = A_i \cap A_j \cap A_k$ and so on, then

$$\left| \bigcup_{i=1}^{n} A_i \right| = \sum_i |A_i| - \sum_{i<j} |A_{ij}| + \sum_{i<j<k} |A_{ijk}| - \cdots \pm |A_{12\ldots n}|.$$

To prove this, consider an element which belongs to exactly r of the A_i. On the left it is counted once. The number of times it is counted on the right is given by

$$\begin{aligned}&\binom{r}{1}-\binom{r}{2}+\binom{r}{3}-\cdots\pm\binom{r}{r}\\=&1-(1-1)^n\\=&1.\end{aligned}$$

Problem

In how many ways can n couples line up such that at least one couple occupies adjacent spaces in the line?

Call the couples $C_1, C_2, \ldots C_n$.

Let A_i be the set of arrangements where couple C_i occupies adjacent spaces, and define A_{ij}, A_{ijk} and so on as above.

The number we seek is precisely the number of elements in the union of the A_i, so we need to find $|A_i|, |A_{ij}|$ and so on.

Suppose that we are told that the couples $C_1, C_2, \ldots C_r$ must each occupy adjacent spaces (we do not care whether the remaining couples do or not). In how many ways can this be accomplished?

We have r couples and $2n-2r$ individuals who are not in the specified couples. This gives $2n-r$ objects which can be arranged in $(2n-r)!$ orders. Once an order for these objects has been chosen, we must choose the order of each couple in one of two ways. Thus $|A_{12\ldots r}| = 2^r(2n-r)!$.

To find, say, $\sum|A_{ij}|$ we note that there are $\binom{n}{2}$ ways to choose the subscripts, and that each A_{ij} has the same number of elements as A_{12}.

Putting all this together with the PIE gives us a final answer of

$$\binom{n}{1}2^1(2n-1)!-\binom{n}{2}2^2(2n-2)!+\binom{n}{3}2^3(2n-3)!-\cdots\pm\binom{n}{n}2^n(2n-n)!.$$

This is not a particularly pretty expression, but the problem is solved.

Generating functions

Given a sequence $a_0, a_1, a_2, \ldots$ we define the *generating function* of the sequence as $G(x) = a_0 + a_1x + a_2x^2 + \cdots + a_kx^k + \ldots$.

It is possible to view generating functions either as *formal power series*, which are never evaluated at any particular value of x, or to view them as genuine functions, in which case care must be taken to determine which values of x cause the infinite series to converge to a limit. Making either approach fully rigorous is fairly technical and is beyond the scope of this book.

The combinatorial interest of generating functions is illustrated in the following example. Consider the sequence $0, 1, 5, 19, 65, \ldots$ defined by

$$a_0 = 0, a_1 = 1;$$
$$a_{n+1} = 5a_n - 6a_{n-1}.$$

We have

$$G(x) = x + 5x^2 + 19x^3 + 65x^4 + \ldots$$
$$xG(x) = x^2 + 5x^3 + 19x^4 + \ldots$$
$$x^2G(x) = x^3 + 5x^4 + \ldots$$

So $G(x) - 5xG(x) + 6x^2G(x) = x$ since all the other terms cancel by the definition of the sequence. Solving for $G(x)$ and using partial fractions we obtain

$$\begin{aligned} G(x) &= \frac{x}{1 - 5x + 6x^2} \\ &= \frac{x}{(1-2x)(1-3x)} \\ &= \frac{1}{1-3x} - \frac{1}{1-2x}. \end{aligned}$$

Each term in the last line can be viewed as the sum of a geometric series so

$$G(x) = (1 + 3x + (3x)^2 + (3x)^3 + \ldots) - (1 + 2x + (2x)^2 + (2x)^3 + \ldots).$$

Collecting like terms we see that

$$G(x) = x(3-2) + x^2(3^2 - 2^2) + x^3(3^3 - 2^3) + \ldots.$$

The definition of $G(x)$ now shows that $a_n = 3^n - 2^n$ so we have used a generating function to solve a recurrence relation.

Induction

Suppose we wish to prove that a claim is true for all integers n greater or equal to some specific positive integer n_0 (often 1).

One way to prove a statement of this form is called *proof by induction.*

First we prove that the claim is true for n_0. This is called checking the *base case*. Then we prove that *if* the claim is true for some number k, *then* it must also be true for the next number, $k+1$. This is called the *inductive step*.

Now you reason as follows: the claim is true for the base case, so the induction step means it must also be true for the next number. This, however, means it must also be true for the number after that one, and the number after that one and so on. This chain never stops, and so we conclude that the claim must be true for all $n \geq n_0$.

The following example is well-known and serves to illustrate the structure of an induction proof.

Problem

Prove that, for all $n \geq 1$, we have

$$1+2+3+\cdots+n = \frac{1}{2}n(n+1).$$

We proceed by induction on n.

Base $(n=1)$.

When $n=1$ we are required to prove that $1 = \frac{1}{2}(1+1)$ which is true.

Step

We assume that $1+2+3+\cdots+k = \frac{1}{2}k(k+1)$ for some k.

Now we consider the sum $1+2+3+\cdots+k+(k+1)$.

Since we assumed our claim was true when $n=k$, we may write

$$[1+2+3+\cdots+k]+(k+1) = \frac{1}{2}k(k+1)+(k+1)$$

which simplifies to $\frac{1}{2}(k+1)(k+2)$.

This is what we were asked to prove, so, by induction, the claim holds for all $n \geq 1$.

In the previous example the claim we proved had the form:

'For every $n \geq 1$, *such and such* is true for n.'

However, it is sometimes useful to prove a claim of the form:

'For every $n \geq n_0$, *such and such* is true for all positive integers less than or equal to n.'

This amounts to the same thing, but allows us to assume slightly more in our induction step, since we can assume the claim holds, not only for k but for smaller numbers as well.

This style of induction is called *complete induction* (or *strong induction*). It is indispensable for the following problem.

Problem

Prove that every polygon can be divided into triangles using non-crossing diagonals.

This result is obvious if the polygon does not have any dents or spikes. However, there is the possibility that if we start drawing diagonals on a sufficiently spiky polygon, we may reach a situation where every remaining diagonal crosses either an edge of the polygon, or one of the diagonals we have already drawn. Our task is to prove that this can always be avoided.

We proceed by induction on n, the number of vertices of the polygon.

Base $(n = 3)$.

When $n = 3$ we have a triangle already, so there is nothing to prove.

Step

Suppose the result holds for all $n \leq k$, and consider a polygon Π with $k + 1$ vertices.

We aim to find a diagonal of which does not cross any sides of Π.

Let X, Y and Z be consecutive vertices of Π. If the segment XZ does not cross any sides of Π, then we may take this segment as our diagonal.

Otherwise, there are other vertices of Π inside triangle XYZ. Now consider a line through Y parallel to XZ and imagine sliding it towards XZ. This line will meet another vertex of Π inside triangle XYZ before it reaches XZ. If the first vertex it meets is called W, then YW cannot cross any other sides of Π. An example is shown in figure 6.16. (If the line hits more than one vertex at the same time, then any one of them can be called used for W.)

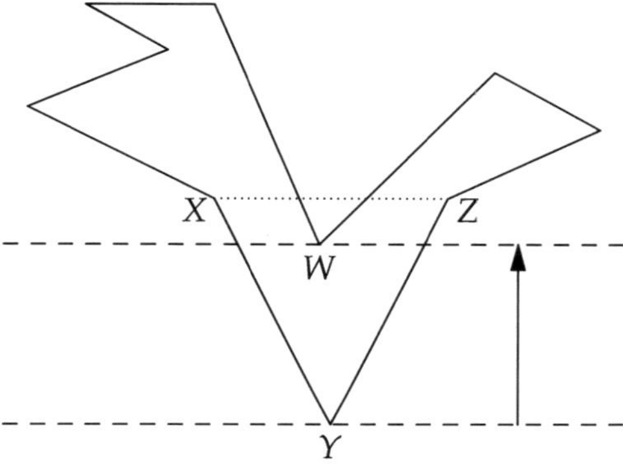

Figure 6.16

Having found this diagonal the rest is easy. The diagonal divides Π into two polygons, each with at most k vertices. Our inductive hypothesis means that each of these can be triangulated using non-crossing diagonals, which gives the required triangulation of Π.

Solutions

Exercise 1a

1. (a) The total number of arrangements is $(2n)!$. For $1 \leq k \leq n$, let A_k be the number of arrangements in which k specified couples are sitting together with the wife to the right of the husband. This means that the remaining $2n - 2k$ people, plus the k couples, are permuted, and thus $\mid A_k \mid = (2n-k)!$. Note that this formula also works when $k = 0$. Hence, using PIE, the number of arrangements is given by

$$\alpha(n) = \sum_{k=0}^{n} (-1)^k \binom{n}{k} (2n-k)!.$$

Then $\alpha(6*) = 2\,872\,050\,480$.

(b) The only difference between this and the previous case is that each couple can be arranged in two ways. We therefore have $\mid A_k \mid = 2^k(2n-k)!$ and

$$\beta(n) = \sum_{k=0}^{n} (-1)^k \binom{n}{k} 2^k (2n-k)!.$$

We have $\beta(6) = 168\,422\,400$.

(c) Begin by seating the husbands; this can be done in $n!$ ways. Next place the wife of the left most husband; there is only one seat for her. Now deal with the second-left husband; she can go in any of three places. Continuing the process, we obtain

$$\gamma(n) = 1 \times 3 \times 5 \times \cdots \times (2n-1)n! = \frac{(2n)!}{2^n}$$

and so $\gamma(6) = 7\,484\,400$.

2. (a) Now the total number of arrangements is $2 \times (n!)^2$, since the row can begin with either a man or a woman and then we permute amongst each group of n people.
Let A_k (for $1 \leq k \leq n-1$) be the number of ways in which k specified couples are sitting together with the wife to the left of the husband.
If the row begins with a woman, we begin by placing the $n-k$ individual men and $n-k$ individual women, which can be done in $(n-k)!^2$ ways. There are now $n-k+1$ gaps, into which k markers are placed; this can be done in $\left(\!\binom{n-k+1}{k}\!\right)$ ways, and then the couples are assigned to the markers in $k!$ ways.
If the row begins with a man, the argument is the same, except that there are only $n-k$ gaps in which to place the markers.
If $k = n$, however, the argument is different. The row consists simply of n couples with the woman on the left, and there are $n!$ arrangements.
Hence $A_k = \left\{\left(\!\binom{n-k+1}{k}\!\right) + \left(\!\binom{n-k}{k}\!\right)\right\}k!(n-k)!^2$. Now we can simplify the two combinations without replacement to obtain $A_k = \left\{\binom{n}{k} + \binom{n-1}{k}\right\}k!(n-k)!^2$.
Now PIE gives

$$\delta(n) = \sum_{k=0}^{n-1}(-1)^k\binom{n}{k}\left\{\binom{n}{k} + \binom{n-1}{k}\right\}k!(n-k)!^2 + (-1)^n n!$$

and we have $\delta(6) = 413\,280$.

(b) This is easier than the previous exercise since the problem is symmetrical. The row begins either M or W but the argument is the same for both. Let A_k (for $1 \leq k \leq n-1$) be the number of ways in which k specified couples are sitting together. Begin by placing the individuals: there are $(n-k)!^2$ ways to do this.

Now insert k markers into the gaps, allowing more than one in a gap; there are $\left(\!\binom{2(n-k)+1}{k}\!\right)$ to do this. Finally allocate the couples to markers in $k!$ ways. Hence $A_k = 2(n-k)!^2\binom{2n-k}{k}k!$ Now PIE gives

$$\begin{aligned}\epsilon(n) &= 2\sum_{k=0}^{n}(-1)^k\binom{n}{k}(n-k)!^2\binom{2n-k}{k}k! \\ &= 2n!\sum_{k=0}^{n}(-1)^k(n-k)!\binom{2n-k}{k}\end{aligned}$$

and so $\epsilon(6) = 138\,240$.

(c) First place the husbands in $n!$ ways. Now work from the left as before, and it becomes apparent that there is only way of placing their wives. So $\zeta(n) = n!$ and $\zeta(6) = 720$.

3. First we count the number of arrangements with the k couples $C_1, C_2, \ldots C_k$ sitting opposite each other. (The analysis will work even if $k = 0$.)

We start by dividing the $2n - 2k$ people not in the specified couples into arbitrary pairs. This can be done in $\dfrac{(2n-2k)!}{2^{n-k}(n-k)!}$ ways. Now we have n pairs, which can be seated in $\dfrac{n! \times 2^n}{2n} = (n-1)!2^{n-1}$ ways. Now PIE gives

$$\psi(n) = \sum_{k=0}^{n}(-1)^k\binom{n}{k}\frac{(2n-2k)!}{2^{n-k}(n-k)!}(n-1)!2^{n-1}.$$

This can be simplified to

$$\psi(n) = \frac{1}{2n}\sum_{k=0}^{n}(-2)^k\binom{n}{k}^2 k!(2n-2k)!$$

and we have $\psi(6) = 23\,193\,600$.

Exercise 1b

1. (a) If (B, e) is one pair, then the others are (C, h), (D, s), (F, i), (G, j) and (W, a). If, however, (B, a) is a pair, the rest are (C, e), (D, h), (F, s), (G, i) and (W, j). So there are precisely two matchings.

   (b) Despite the apparent prodigality of the gentlemen's affections, there cannot be a matching here since all six men together have chosen only five ladies.

   (c) The gentlemen and ladies split into two groups of three. Once B has chosen between a and h, C and D have no options, so there are two ways of pairing these three men. The same thing happens with F, G and W so there are four matchings altogether.

   (d) There are 24 ways of satisfying B, C, D and F and a further 12 of pairing G and W so there are 288 matchings.

   (e) This is impossible, since the pool of B, D, F, G and W is of size four.

2. (a) A possible matching is (B, j), (C, h), (D, a), (F, l), (G, m), (W, e)

   (b) The three sisters are popular, but a matching is impossible because the pool of all six gentlemen only contains five ladies.

   (c) This is impossible. The pool of B, D, F, G and W omits both j and l, and there is only one gentlemen left for these two ladies.

3. Let $B = \{g_{N+1}, g_{N+2}, \ldots, g_M\}$ be the set of bogus gentlemen. We shall show that there is a matching for $G \cup B$ and L.

   If we can show that any subset of k gentlemen in $G \cup B$ has a pool consisting of at least k ladies, then Hall's theorem guarantees that there is a matching. If we happen to choose a subset consisting entirely of the original gentlemen, we know by condition (a) that this is the case. So the only interesting case is when the subset chosen contains *at least one* of the bogus gentlemen. Suppose it consists of x real and $k - x$ bogus gentlemen, with $k > x$.

   At this point in the argument, a couple of Venn diagrams turn out to be useful.

   Figure 1b.A shows the N real and the $M - N$ bogus gentlemen, with a subset of size k consisting of x who are real and $k - x$ who are bogus.

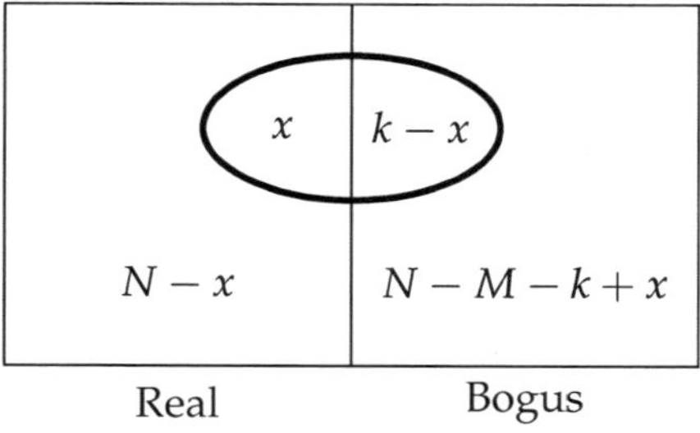

Figure 1b.A

Figure 1b.B shows the M ladies, of whom m are sisters. The subset shown represents the pool of the *real* gentlemen, who have chosen a sisters and b others. Recall that the pool of the bogus gentlemen is the $M - m$ other ladies.

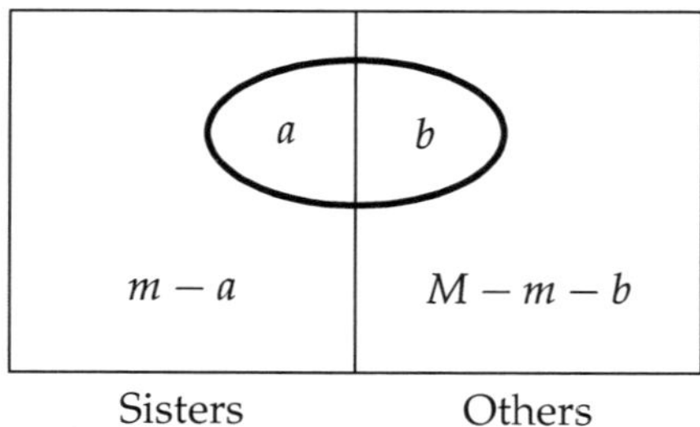

Figure 1b.B

We apply condition (ii) to the x real gentlemen to obtain the fact that there are at most $N - x$ sisters *not* in their pool, so $m - a \leq N - x$. Now we count the number of ladies in the pool for all the gentlemen, real and bogus. It contains $a + (M - m)$ ladies and we have that $a + M - n \geq M - N + x > k$. This, by Hall's theorem, guarantees that we have a matching for all of the gentlemen.

We then transport all the bogus gentlemen, along with their consorts, who are not in S, to the colonies. Those couples who remain on home turf include all the sisters successfully paired with the real gentlemen.

4. The set X is the 52 cards, consisting of 13 of each denomination from A to 2. The piles are thirteen sets of four cards represented by

their denominations. For example, a pile consisting of the Ace of Spades, 10 of Spades, 10 of Hearts and 5 of Clubs will represented by $\{A, 10, 5\}$.

Between them k piles contain exactly $4k$ cards, and, as there are only 4 cards for a particular denomination, they will contain at least k denominations. Hence the condition for a transversal is achieved, and it will consist of 13 cards, each of a different denomination.

5. The condition that every subset of k boys collectively knows at least k girls (from the boys' viewpoint) is sufficient for a matching to exist. If there is a matching for the boys, there is also a matching for the girls. Hence the condition (from the girls' viewpoint) is necessary, and so every subset of k girls knows at least k boys.

6. The set X is the $2n$ players, and A_i, for $1 \leq i \leq 2n-1$, is the set of winners on day i. If, after k days, there is anybody who has lost all of their matches, they will have lost to k different people, so the number who have won at least one match will be greater than or equal to k. If nobody has lost all their matches, then the number who have won at least one match will be $2n$. In either case, we see that, after k days, $\left|\bigcup_{1\leq i\leq k} A_i\right| \geq k$. Hence, by Hall's theorem, we can pick a transversal, which consists of a list of $2n-1$ different winners as required.

7. In the language of the marriage theorem, we can think of the columns as gentlemen and the rows as ladies. If there is a rook in cell (i, j) we will say that gentlemen i would consider marrying lady j.

   The set of $2n$ rooks we hope to find (one in each row and column) represents a matching between the ladies and the gentlemen. Thus we will be done by Hall's theorem, provided every set of k gentlemen has a pool of at least k ladies.

   Suppose, for contradiction, that there is a set of k columns such that the nk rooks in these columns lie in fewer than k different rows. This would imply that the average number of rooks in these rows is greater than n, which is impossible since every row contains exactly n rooks.

8. Consider, for each column of the rectangle, the set of choices we have for the next element. We have n sets each containing $n-r$ elements, since the elements in each column of the rectangle are all different.

Note that any element, from 1 to n, occurs in exactly $n-r$ of the sets, since it is entered into a different column for each row of the rectangle.

When we consider any $n-r$ of these sets and make a list of all their elements, it will contain repetitions. However, no element can be in this list more than $n-r$ times, so, by the pigeonhole principle, the list must contain at least $n-r$ different elements. Therefore the union of the these sets (which does not allow repetitions) contains at least $n-r$ elements. Hence, by the marriage theorem, a transversal exists, and this is exactly what we need to build another row below the rectangle. Continuing this process until there are n rows, we obtain an $n \times n$ Latin square.

9. This is impossible. In fact, we cannot even extend the given array to a Latin 3×7 rectangle, since we would need a 3 in each of the two new columns, so there would be a column with two 3s in it.

Exercise 1c

1. There are seven essentially different solutions (which cannot be transformed into one another by renaming the schoolgirls). The list shown in figure 1c.A on the following page is organised in such a way that each list resembles the previous one, up to a point.

2. (a) The solution is shown in table 1c.A.

$\{1,6\}$	$\{5,6\}$	$\{4,6\}$	$\{3,6\}$	$\{2,6\}$
$\{2,5\}$	$\{1,4\}$	$\{5,3\}$	$\{4,2\}$	$\{3,1\}$
$\{3,4\}$	$\{2,3\}$	$\{1,2\}$	$\{5,1\}$	$\{4,5\}$

Table 1c.A

   (b) The two are equivalent, as can be seen by interchanging 2 with 6 and 3 with 4.

Exercise 1d

1. The result is another Fano plane, with the points relabelled as shown in figure 1d.A. The same triples of points as before form each line,

{1,2,3}	{1,4,7}	{1,5,14}	{1,9,13}	{1,8,10}	{1,6,12}	{1,11,15}
{4,5,6}	{2,5,10}	{2,4,15}	{2,7,12}	{2,11,13}	{2,9,14}	{2,6,8}
{7,8,9}	{3,6,13}	{3,8,12}	{3,4,11}	{3,5,9}	{3,10,15}	{3,7,14}
{10,11,12}	{8,11,14}	{6,9,11}	{5,8,15}	{4,12,14}	{4,8,13}	{4,9,10}
{13,14,15}	{9,12,15}	{7,10,13}	{6,10,14}	{6,7,15}	{5,7,11}	{5,12,13}
{1,2,3}	{1,4,7}	{1,5,14}	{1,12,13}	{1,9,10}	{1,6,8}	{1,11,15}
{4,5,6}	{2,5,10}	{2,4,15}	{2,7,11}	{2,8,13}	{2,9,14}	{2,6,12}
{7,8,9}	{3,6,13}	{3,8,12}	{3,4,9}	{3,5,11}	{3,10,15}	{3,7,14}
{10,11,12}	{8,11,14}	{6,9,11}	{5,8,15}	{4,12,14}	{4,11,13}	{4,8,10}
{13,14,15}	{9,12,15}	{7,10,13}	{6,10,14}	{6,7,15}	{5,7,12}	{5,9,13}
{1,2,3}	{1,4,7}	{1,5,14}	{1,8,13}	{1,11,15}	{1,6,12}	{1,9,10}
{4,5,6}	{2,5,10}	{2,4,15}	{2,9,14}	{2,12,13}	{2,10,11}	{2,6,8}
{7,8,9}	{3,6,13}	{3,8,12}	{3,4,11}	{3,5,9}	{3,10,14}	{3,7,15}
{10,11,12}	{8,11,14}	{6,9,11}	{5,7,11}	{4,8,10}	{4,9,13}	{4,12,14}
{13,14,15}	{9,12,15}	{7,10,13}	{6,10,15}	{6,7,14}	{5,8,15}	{5,11,13}
{1,2,3}	{1,4,7}	{1,5,14}	{1,9,10}	{1,8,13}	{1,6,12}	{1,11,15}
{4,5,6}	{2,5,10}	{2,4,15}	{2,12,13}	{2,7,11}	{2,9,14}	{2,6,8}
{7,8,9}	{3,6,13}	{3,8,12}	{3,4,11}	{3,5,9}	{3,7,15}	{3,10,14}
{10,11,12}	{8,11,14}	{6,9,11}	{5,8,15}	{4,12,14}	{4,8,10}	{4,9,13}
{13,14,15}	{9,12,15}	{7,10,13}	{6,7,14}	{6,10,15}	{5,11,13}	{5,7,12}
{1,2,3}	{1,4,7}	{1,5,15}	{1,9,14}	{1,8,10}	{1,6,12}	{1,11,13}
{4,5,6}	{2,5,10}	{2,4,14}	{2,7,12}	{2,11,15}	{2,9,13}	{2,6,8}
{7,8,9}	{3,6,13}	{3,8,12}	{3,4,11}	{3,5,9}	{3,10,14}	{3,7,15}
{10,11,12}	{8,11,14}	{6,9,11}	{5,8,13}	{4,12,13}	{4,8,15}	{4,9,10}
{13,14,15}	{9,12,15}	{7,10,13}	{6,10,15}	{6,7,14}	{5,7,11}	{5,12,14}
{1,2,3}	{1,4,7}	{1,5,15}	{1,12,14}	{1,9,10}	{1,6,8}	{1,11,13}
{4,5,6}	{2,5,10}	{2,4,14}	{2,7,11}	{2,8,15}	{2,9,13}	{2,6,12}
{7,8,9}	{3,6,13}	{3,8,12}	{3,4,9}	{3,5,11}	{3,10,14}	{3,7,15}
{10,11,12}	{8,11,14}	{6,9,11}	{5,8,13}	{4,12,13}	{4,11,15}	{4,8,10}
{13,14,15}	{9,12,15}	{7,10,13}	{6,10,15}	{6,7,14}	{5,7,12}	{5,9,14}
{1,2,3}	{1,4,7}	{1,5,12}	{1,6,15}	{1,8,14}	{1,9,10}	{1,11,13}
{4,5,6}	{2,5,10}	{2,4,14}	{2,12,13}	{2,9,11}	{2,6,8}	{2,7,15}
{7,8,9}	{3,8,13}	{3,7,11}	{3,4,9}	{3,5,15}	{3,12,14}	{3,6,10}
{10,11,12}	{6,11,14}	{6,9,13}	{5,8,11}	{4,10,13}	{4,11,15}	{4,8,12}
{13,14,15}	{9,12,15}	{8,10,15}	{7,10,14}	{6,7,12}	{5,7,13}	{5,9,14}

Figure 1c.A

and each point has the same three lines through it. Hence it can be considered as the same plane.

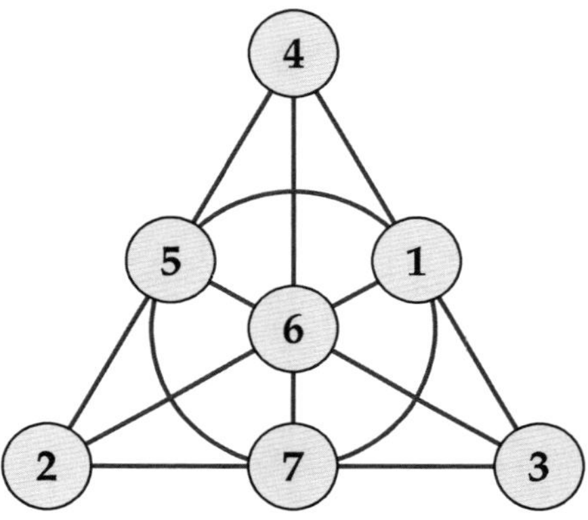

Figure 1d.A

2. One way of creating the correspondence is given below. It produces the planes shown in figure 1d.B. It is easier to check the required property, which preserves *incidence*.

$$1 \leftrightarrow 237 \qquad 2 \leftrightarrow 134 \qquad 3 \leftrightarrow 126 \qquad 4 \leftrightarrow 245$$
$$5 \leftrightarrow 467 \qquad 6 \leftrightarrow 356 \qquad 7 \leftrightarrow 157$$

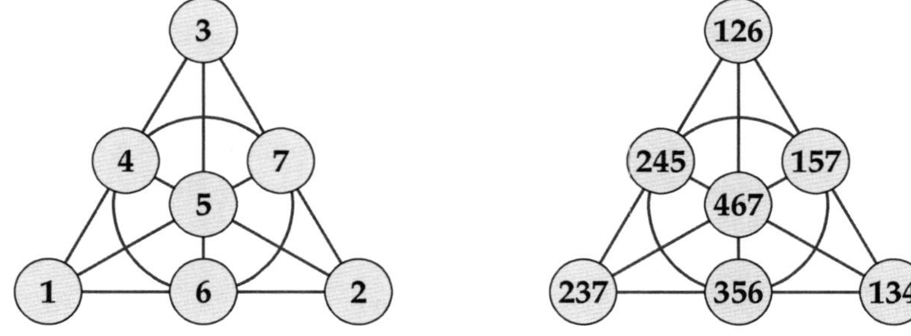

Figure 1d.B

These two exercises show that the Fano plane is intensely democratic; points and lines are interchangeable, and every point or line has the same properties as any other.

3. (a) There are $\binom{7}{3} - 7 = 28$ circles and $\binom{6}{2} - 3 = 12$ through each point. Note that $28 \times 3 = 12 \times 7$.

   (b) At any point there are three lines, but two of these will contain two points of the circle. The third line is the only tangent to the circle.

4. One way of doing this is shown in figure 1d.C.

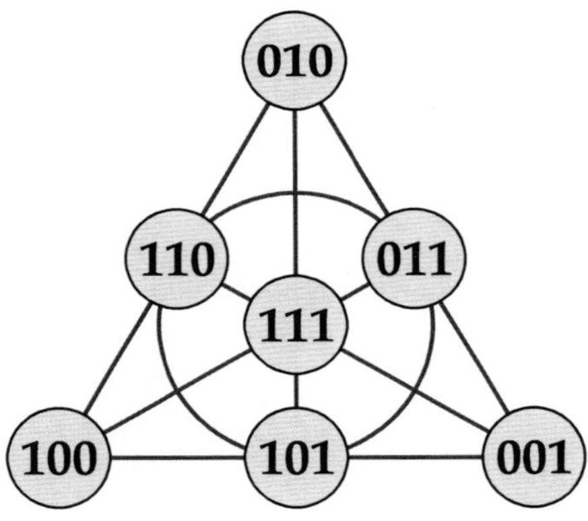

Figure 1d.C

5. (a) This is immediate from the third bullet point in the definition.

   (b) Consider two lines meeting in a point. Each line must contain two points other than the intersection. Split these into two pairs and draw two more lines meeting in a point P. Now you can set up a projection from P mapping the points on one line to the points on any other, establishing a one-to-one correspondence. Similarly, from any point in the plane, there is a one-to-one correspondence between lines through it and points on a line. It follows that every line has the same number of points and every point lies on the same number of lines, and there are the same number of points and lines.

(c) Take an arbitrary line l and an arbitrary point P on it. There are n other lines through P, each containing n points other than P, and every point on the plane, apart from those on l, is accounted for in this way. Hence there are n^2 points not on l and so, together with the points on l, there are $n^2 + n + 1$ points in the plane. By (b), there are also $n^2 + n + 1$ lines.

(d) There must be thirteen points and thirteen lines. Figure 1.24 on page 32 describes a geometry with points labelled 1 to 9 and twelve lines, each of which contains three points. The figure is reproduced below as figure 1d.D. We wish to extend this to a geometry with thirteen points and thirteen lines, each of which contains four points.

1	2	3
4	5	6
7	8	9

1	2	3
4	5	6
7	8	9

1	2	3
4	5	6
7	8	9

1	2	3
4	5	6
7	8	9

$$\begin{array}{cccc} \{1,2,3\} & \{1,4,7\} & \{1,5,9\} & \{1,6,8\} \\ \{4,5,6\} & \{2,5,8\} & \{2,6,7\} & \{2,4,9\} \\ \{7,8,9\} & \{3,6,9\} & \{3,4,8\} & \{3,5,7\} \end{array}$$

Figure 1d.D

We add the points A, B, C and D which we say form a line. Now we need to add a letter to each of the triples of numbers such that any pair of lines meets in a single point.
The quadruples shown in figure 1d.E have the required property.

$$\{A, B, C, D\}$$

$$\begin{array}{cccc} \{1,2,3,A\} & \{1,4,7,B\} & \{1,5,9,C\} & \{1,6,8,D\} \\ \{4,5,6,A\} & \{2,5,8,B\} & \{2,6,7,C\} & \{2,4,9,D\} \\ \{7,8,9,A\} & \{3,6,9,B\} & \{3,4,8,C\} & \{3,5,7,D\} \end{array}$$

Figure 1d.E

(e) Dobble is clearly associated with a finite projective plane of order 7, with the symbols corresponding to points and the cards to lines (or vice versa). This, however, has $49 + 7 + 1 = 57$ points and 57 lines. In the game, two cards are missing. Since these cards would each contain 7 symbols, with one in common, the symbols do not appear equally frequently in the pack. It turns out to be the Snowman which is the symbol common to the two missing cards, and consequently there are only 6 Snowmen in the pack. Another seven symbols only appear 7 times. It is not difficult to identity two more cards which would complete the projective plane. In a full pack, every pair of symbols would appear on a unique card. Were the game to consist of a full pack, this property could be exploited, requiring players to identify, say, the card with a Shark and a Sun on it.

It is not clear why the game's publishers chose to omit two cards. One useful consequence of this decision is that if a single *starting card* is chosen, then the remaining cards can be equally divided among either two or three players.

Exercise 2a

1. If each vertex is joined to every other vertex, then there are n vertices of degree $(n-1)$. Thus, by the handshaking lemma, $|E| = \frac{n(n-1)}{2}$. Alternatively, we can specify an edge by choosing two vertices, so the number of possible edges is $\binom{n}{2}$.

2. Possible drawings of the eleven graphs are shown in figure 2a.A.

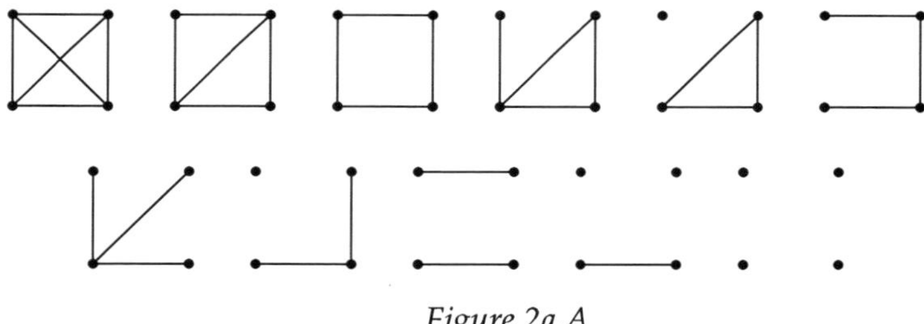

Figure 2a.A

3. Possible drawings of the six graphs are shown in figure 2a.B.

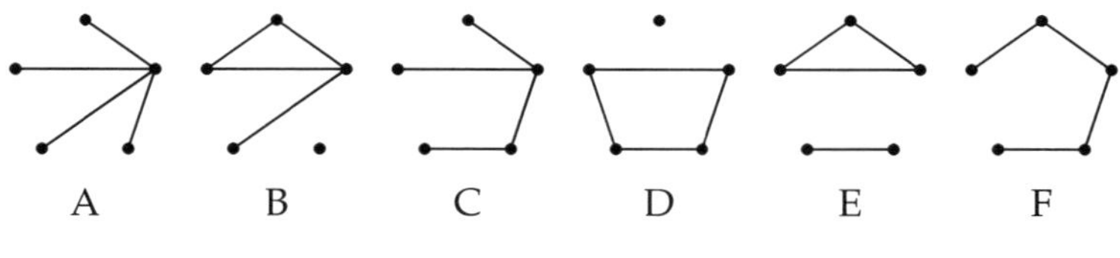

Figure 2a.B

Showing that this list contains all the graphs with four edges on five vertices requires some care. If there is a vertex of degree four, we have graph A. If there is a vertex v of degree three, then we have one more edge to place. If it joins two vertices both of which are joined to v we have graph B, and otherwise we have graph C. If the highest degree is two, then the degree sequence must be $2, 2, 2, 2, 0$ or $2, 2, 2, 1, 1$. The first case gives us graph D and the second gives us E and F.

4. No. The sum of the degrees is not even which contradicts the handshaking lemma.

5. No. There are only six numbers in the sequence, and the maximum degree of any vertex in a graph on six vertices is five.

6. Yes. One possible graph is shown in figure 2a.C.

Figure 2a.C

7. No. There are eight numbers in the sequence, so the maximum degree is seven. However, if a graph on eight vertices has a vertex of degree seven, then it cannot have a vertex of degree zero. The same argument shows that in any graph there are two vertices with the same degree.

8. Yes. The vertices and edges of a cube form one such graph, though there are many possible graphs.

9. The two possible graphs are shown in figure 2a.D.

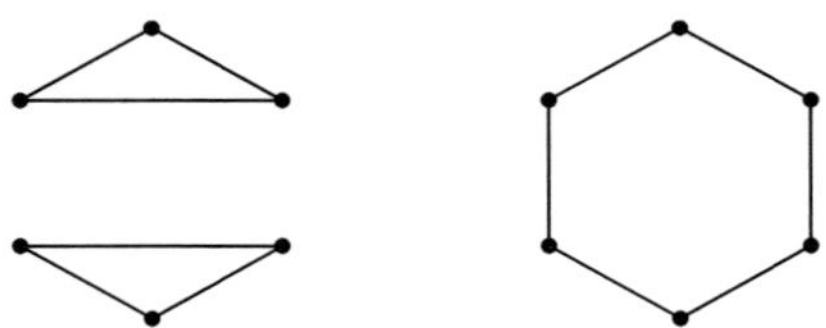

Figure 2a.D

10. Suppose that there is a graph G with an odd number of vertices, an even number of which are even. The sum of all the degrees in the G is odd. This contradicts the handshaking lemma, so there is no such graph.

Exercise 2b

1. The six graphs are shown in figure 2b.A.

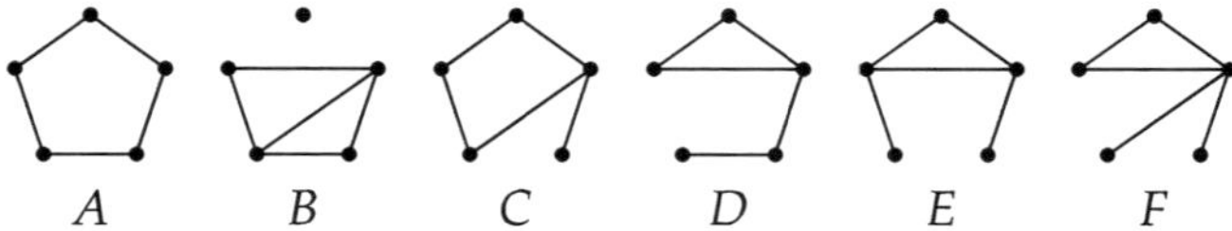

Figure 2b.A

If a graph with five vertices contains more than four edges, then it cannot be a tree, or be made up of components which are themselves trees. Therefore it contains a cycle. We list the graphs systematically by the length of the longest cycle they contain.

A 5-cycle gives us graph A.

If the graph contains a 4-cycle, then there is one more edge to place. If this edge joins two vertices in the 4-cycle we have graph B, otherwise we have graph C.

If the graph contains a 3-cycle, then there are two other vertices not in the 3-cycle. If these are joined to each other then we have graph D. If each of them is joined to a different vertex in the 3-cycle we have E and if both are joined to the same vertex in the 3-cycle we have F.

2. One such graph is shown in figure 2b.B.

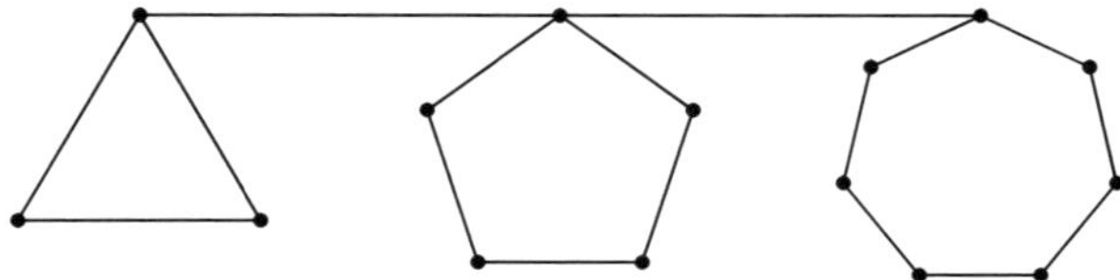

Figure 2b.B

3. It is given that G contains no cycles, so it remains to check that G is connected. Suppose that G is not connected. Adding an edge between two vertices in different components does not add any cycles, which contradicts the definition of G. Therefore, G is connected and thus a tree.

4. Suppose G is a tree. We form the reduced graph G' by removing an edge $\{v, w\}$ from G. If there were a path from v to w in G', then adding $\{v, w\}$ would form a cycle in G which is impossible.

Therefore, v and W are in different components of G', so $\{v, w\}$ is a bridge.

For the converse, suppose that G is a connected graph where every edge is a bridge. Removing an edge from a cycle cannot disconnect a graph, so G contains no cycles. Thus G is a tree as required.

5. Consider the set of connected subgraphs of G which contain all vertices in G. This set is not empty since it contains G itself.

   Let T be a member of this set which contains the smallest possible number of edges. If T contained an edge which were not a bridge, that edge could be removed to form a connected subgraph with fewer edges. Thus every edge of T is a bridge, so T is a tree.

6. Suppose a graph with no odd vertices contained a bridge. Removing this bridge would form two components each with precisely one odd vertex. However, a component cannot have an odd number of odd vertices as this would contradict the handshaking lemma.

7. Yes. Consider the degree sequence 2, 2, 2, 1, 1. This is the degree sequence of a path with four edges, which is a tree. It is also the degree sequence of a 3-cycle and an isolated edge. These are graphs E and F in the solution to question 3 in exercise 2a.

8. We can rephrase the question as follows. What is the smallest number of edges which must be removed from the complete graph K_n in order to disconnect it. Once the graph is disconnected, its vertices can be divided into two sets, with no edges between them. If the sets have size k and $n - k$, then at least $k(n - k)$ edges must have been removed. For $1 \leq k \leq n - 1$ the quantity $k(n - k)$ is at least $n - 1$ (why?), so a disconnected graph cannot have more than $\frac{(n-1)(n-2)}{2}$ edges. This bound is attained by a copy of K_{n-1} and an isolated vertex.

9. (a) Suppose that G is a bipartite graph and that it has an odd cycle with vertices $v_1 v_2 \dots v_{2k+1}$. We call the vertex classes A and B and assume that $v_1 \in A$. This implies $v_2 \in B$ which in turn shows $v_3 \in A$. Continuing in this way, we see that all the odd-numbered vertices belong to A, but this is impossible since $\{v_1, v_{2k+1}\}$ is an edge of G.

   (b) We call the vertex classes A and B and choose a vertex v which we place in A. For each other vertex w we place w in A if, and only if, the shortest path from v to w uses an even number of edges.

We claim that no edge joins two vertices in the same class.

Suppose that G contains the edge $\{x, y\}$ where x and y are in the same class. If we go along the shortest path from v to x, across the edge $\{x, y\}$ and then along the shortest path from y to v we obtain a closed walk with an odd number of edges.

This closed walk may not itself be a cycle, but we can prove by (complete) induction that an odd closed walk contains or is an odd cycle.

The base case is the odd walk of length three which is a 3-cycle.

For the inductive step, we assume that all odd walks of length less than $2k$ contain an odd cycle and consider an odd closed walk whose vertices are $v_1 v_2 \ldots v_{2k+1} v_1$. If there are no repeated vertices apart from the first and last v_1, then the walk is a cycle. Otherwise we may break the walk into two shorter closed walks one of which must be odd.

This shorter odd walk contains an odd cycle by induction so we are done.

Exercise 2c

1. Any drawing of K_5 contains a copy of K_4. That is, it shows four vertices with every possible edge between them. If these six edges do not cross then they 'look like' figure 2.7. In particular, they divide the plane into four regions of which three are finite and one infinite. Whichever region the fifth vertex is in, there is a vertex it cannot be joined to.

2. Yes. One possible drawing is shown in figure 2c.A: the only edge missing is between v and w.

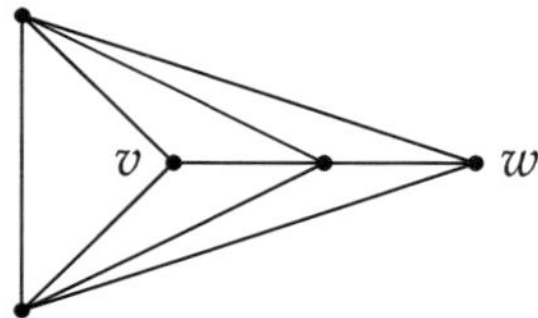

Figure 2c.A

3. Yes. In figure 2c.B the two vertices on the left form one vertex class and those in the horizontal line on the right form the other. This construction can easily be extended to show that $K_{2,n}$ is planar for all n.

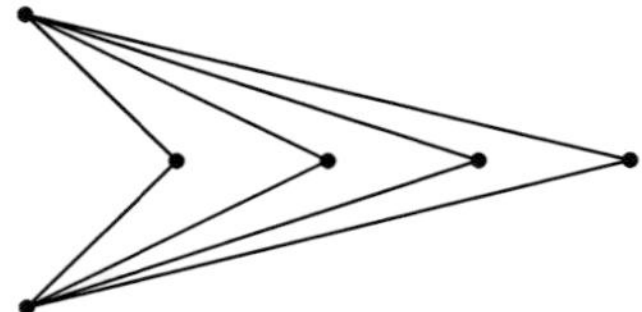

Figure 2c.B

4. No. K_6 contains K_5 as a subgraph so a planar drawing of K_6 would include a planar drawing of K_5 which cannot exist. It also contains $K_{3,3}$ as a subgraph.

5. There are m vertices of degree n and n vertices of degree m. Thus, by the handshaking lemma, there are mn edges.

6. Only K_1, K_2, K_3 and K_4 are planar.

7. The graphs $K_{1,n}$ and $K_{2,n}$ are planar for all n. Every other complete bipartite graph contains $K_{3,3}$ as a subgraph so cannot be planar.

8. In problem 2.12 we showed that

$$\sum_{\mathcal{F} \in F} |\mathcal{F}| = 2e.$$

   Every face has $|\mathcal{F}| \geq 3$ and the result follows.

9. Problem 2.11 showed that $v - e + f = 2$. Thus $3v - 3e + 3f = 6$ which gives the required bound when combined with the previous question.

10. The graph K_5 has $v = 5$ and $e = 10$, so $3v - e = 5 < 6$. Therefore, K_5 is not planar.

11. Question 9 shows that $e \leq 3v - 6 = 15$. To check this bound can be attained we observe that the graph shown in figure 2c.C is planar and has 15 edges. Note that this graph is an equality case for the condition $e \leq 3v - 6$ because all its faces are 3-cycles.

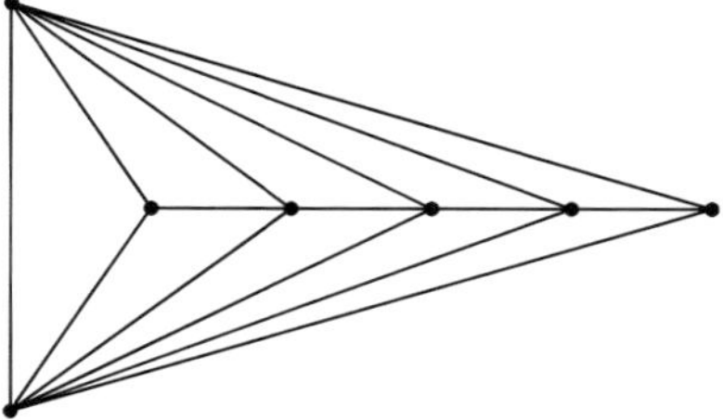

Figure 2c.C

Exercise 2d

1. We will prove that every triangulated n-gon has a 3-colouring by (complete) induction on the number of sides.

   Base: ($n = 3$)

   A 3-cycle can be 3-coloured.

   Step

   Assume the claim holds for all $n \leq k$ and consider an arbitrary triangulated $k + 1$-gon.

   Take any diagonal $\{v, w\}$ and cut the polygon along this diagonal to form two triangulated polygons, each with fewer sides than the original one.

   Each of these polygons can be 3-coloured by induction, so it remains to 'stitch together' the two colourings to form a colouring of the original $(k + 1)$-gon.

   Suppose the 3-colouring on one small polygon assigns colour 1 to v and colour 2 to w. The 3-colouring of the other small polygon also assigns different colours to v and w. However, we can rename the colours in this second colouring so that the colourings agree on v and w. The two colourings can now be combined to form a 3-colouring of the $k + 1$-gon.

   This completes the inductive step.

   An example of the inductive step applied to an irregular 14-gon is shown in figure 2d.A. Colours 2 and 3 need to be exchanged on one of the small polygons before the colourings can be combined.

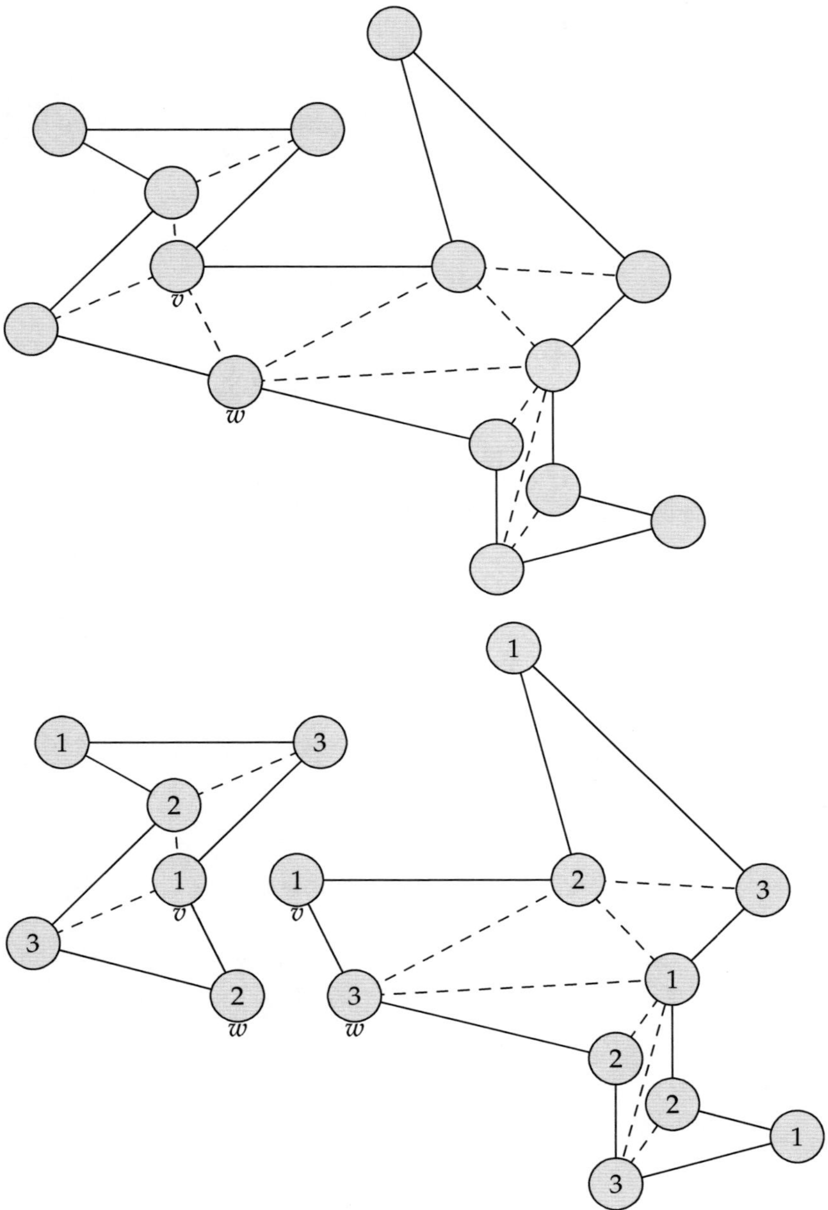

Figure 2d.A

2. We begin by triangulating the gallery and 3-colouring the vertices. Clearly one colour is used at most $\lfloor \frac{n}{3} \rfloor$ times. We call this colour 1. Every triangle in the triangulation has one vertex of each colour, so if we place guards by the colour 1 vertices, then every triangle can be watched over by at least one guard. In particular, every wall is safely guarded.

3. If a planar graph has a face which is not bounded by a 3-cycle, then another edge can be added as a 'diagonal' of that face. Therefore, such a graph cannot be maximal planar.

4. We begin by observing that every vertex in a maximally planar graph on more than three vertices has degree at least three. Indeed, if we remove a vertex x (and the edges which contain it) from the graph we obtain another planar graph. The removed vertex came from a face $\mathcal{F}$ of this reduced graph and must have been joined to every vertex in the walk bounding that face. Therefore $\deg(x) = |\mathcal{F}| \geq 3$.

   Now suppose, for contradiction, that a maximally planar graph on v vertices had at most three vertices with degree 3, 4 or 5. The sum of the degrees in the graph is at least $6(v-3) + 3 \times 3 = 6v - 9$. However, the graph is planar so $3v - e \geq 6$. This implies that the degree sum is at most $6v - 12$. This contradiction completes the proof.

5. We prove the following claim holds for all n.

   If G is a maximally planar graph on n vertices, and a drawing of G using (possibly curved) non-crossing lines has the 3-cycle abc bounding its infinite face, then there is a drawing of G using non-crossing straight lines which also has abc bounding its infinite face.

   Base: ($n \leq 4$)

   The theorem is trivial for graphs on less than five vertices since K_4 can be drawn using non-crossing straight lines, with any 3-cycle bounding the infinite face.

   Step

   Assume the result holds for $n = k$ and consider an arbitrary maximal planar graph G on $k+1$ vertices. Suppose a drawing of G using non-crossing (possibly curved) lines has abc bounding its infinite face, and choose a vertex v with degree at most five which is not one of a, b or c.

Remove vertex v from G to form the reduced graph G'.

By induction, G' can be drawn with non-crossing straight lines and abc bounding the infinite face.

The vertex v was removed from a face $\mathcal{F}$ of G' which has at most five sides by the choice of v.

In the straight line drawing of this graph we may consider the face $\mathcal{F}$ as an art-gallery with at most five walls. This gallery can be guarded by one guard since $\frac{5}{3} < 2$. Therefore, we can place v inside this face, where the guard would be stationed, and join it with straight lines to all of its neighbours. This completes the induction and hence the proof.

6. Given a planar graph, simply add edges until it becomes maximally planar. Now we apply the previous result and then remove the extra edges.

Exercise 3a

1. This and subsequent questions can be solved systematically using tables to help count the equivalence classes. This is done by giving codes to groups of raw colourings.

   Table 3a.A relates to the case where raw colourings are identified under rotations, and table 3a.B extends this to include reflections as well. The three permissible colours are coded by B, G and R.

   The first row of the table relates to colourings using only one colour. This is indicated by the code BBB, but this should be interpreted as covering the cases where all three vertices are green or red. There are 3 such raw colourings, but each equivalence class consists of only one raw colouring, so there are 3 equivalence classes. In the table, the number in the fourth column is divided by the number in the second column.

   The second row of the table is given the code BBG. This represents any situation where two vertices have one colour and the third has a different colour. There are three ways of choosing the colour which is used twice, and two ways of choosing the second colour, and this can occur at any of the three vertices. Hence there are 18 raw colourings. Under rotation, these divide up into equivalence classes of size 3. Hence there are 6 equivalence classes.

   The final row of the table has the code BGR. This refers to any raw colouring where the three vertices have different colours. Hence there are 6 raw colourings, and each equivalence class is of size 3, so there are 2 equivalence classes under this code. Finally we add the third column to give the required 11 equivalence classes. Note, as a check, that the final column adds to 27, the total number of raw colourings.

Raw colouring	Size of class	Classes	Colourings
BBB	1	3	3
BBG	3	6	18
BGR	3	2	6
Totals		11	27

Table 3a.A

If reflections are also allowed, then BGR and BRG belong to the same class, and we obtain table 3a.B.

Raw colouring	Size of class	Classes	Colourings
BBB	1	3	3
BBG	3	6	18
BGR	6	1	6
Totals		10	27

Table 3a.B

It makes no difference whether vertices or sides are coloured, since each vertex is opposite exactly one side.

2. We label raw colourings anticlockwise from the top left corner. For rotations alone, we have table 3a.C. We note that the codes BBGG and BGBG refer to different raw colourings using only two colours. In the first the colours are adjacent in pairs, but in the second the colours alternate.

Raw colouring	Size of class	Classes	Colourings
BBBB	1	3	3
BBBG	4	6	24
BBGG	4	3	12
BGBG	2	3	6
BBGR	4	6	24
BGBR	4	3	12
Totals		24	81

Table 3a.C

If reflections are also permitted, the raw colourings BBGR and BBRG are in the same class, but nothing else changes and we obtain table 3a.D.

Raw colouring	Size of class	Classes	Colourings
BBBB	1	3	3
BBBG	4	6	24
BBGG	4	3	12
BGBG	2	3	6
BBGR	8	3	24
BGBR	4	3	12
Totals		21	81

Table 3a.D

It makes no difference whether vertices or sides are coloured, since we can associate each side with the anticlockwise vertex.

3. Note that the only allowable rotation is a half-turn, since the rectangle is not a square. Again we represent the colour pattern side by side anticlockwise from the top. For such rotations, we construct table 3a.E.

Raw colouring	Size of class	Classes	Colourings
BBBB	1	2	2
BBBR	2	4	8
BBRR	2	2	4
BRBR	1	2	2
Totals		10	16

Table 3a.E

If reflections are also permitted, we obtain table 3a.F.

Raw colouring	Size of class	Classes	Colourings
BBBB	1	2	2
BBBR	2	4	8
BBRR	4	1	4
BRBR	1	2	2
Totals		9	16

Table 3a.F

Exercise 3b

1. The elements are the identity I, rotations R and R^2 of 120° and 240° and reflections A, B and C in the three medians. The group is D_3. It has order six and is shown in table 3b.A.

	I	R	R^2	A	B	C
I	I	R	R^2	A	B	C
R	R	R^2	I	C	A	B
R^2	R^2	I	R	B	C	A
A	A	B	C	I	R^2	R
B	B	C	A	R	I	R^2
C	C	A	B	R^2	R	I

Table 3b.A

If only rotations are allowed, the group is C_3 and consists only of the first three rows and columns.

2. The group consists of the identity I, rotations R, R^2 and R^3 of 90°, 180° and 270° and the following reflections.

   - X in a 'horizontal' axis;
   - Y in a vertical axis;
   - U in the leading diagonal;

- V in the other diagonal.

The group is D_4. It has order 8 and is shown in table 3b.B.

	I	R	R^2	R^3	X	Y	U	V
I	I	R	R^2	R^3	X	Y	U	V
R	R	R^2	R^3	I	V	U	X	Y
R^2	R^2	R^3	I	R	Y	X	U	V
R^3	R^3	I	R	R^2	U	V	Y	X
X	X	U	Y	V	I	R^2	R	R^3
Y	Y	V	X	U	R^2	I	R^3	R
U	U	Y	V	X	R^3	R	I	R^2
V	V	X	U	Y	R	R^3	R^2	I

Table 3b.B

The rotations alone form the group C_4.

3. The group is D_5, of order 10, and consists of the identity, five rotations which are multiples of 36° and five reflections in the axes of symmetry.

4. The group consists of the identity, rotations of 120° and 240° around the four lines from a vertex perpendicular to the opposite face and rotations of 180° around the three lines joining midpoints of opposite edges. It has order 12.

Exercise 3c

1. Question 2: For rotations alone, the group is C_4. We have $|\Omega_I| = 81$. Also $|\Omega_R| = |\Omega_{R^3}| = 3$ since the only unaffected colourings are monochromatic, and $|\Omega_{R^2}| = 9$ since opposite corners must be coloured the same. Hence the number of colourings is indeed $\frac{81+3+9+3}{4} = 24$.

   When reflections are permitted, the group is D_4. We have that $|\Omega_U| = |\Omega_V| = 27$ since the reflections fix colourings with one pair of opposite vertices the same colour and the other two any colour. Also $|\Omega_X| = |\Omega_Y| = 9$ since two pairs of adjacent vertices have the same colour. Hence the number of colourings is $\frac{96+27+27+9+9}{8} = 21$.

Question 3: For rotations alone, the group is C_2. We have $|\Omega_I| = 16$ and $|\Omega_R| = 4$, so then number of colourings is $\frac{16+4}{2} = 10$.

For both rotations and reflections, the group is D_2 . We still have $|\Omega_I| = 16$ and $|\Omega_R| = 4$ and also $|\Omega_X| = |\Omega_Y| = 8$, so the number of colourings is $\frac{16+4+8+8}{4} = 9$.

2. Treat the necklace as an octagon. There are $\binom{8}{4} = 70$ raw colourings and the group is D_8. The invariant set for I is all the colourings so $\Omega_I = 81$. For the rotations R, R^3, R^5 and R_7 the invariant sets are empty. For R_2 and R_6 the invariant sets consist of colourings where the red and blue beads alternate, so they have two members. For R_4 the invariant set consists of colourings with opposite beads of the same colour, so $|\Omega_{R^4}| = \binom{4}{2} = 6$. For the four reflections whose axes pass through two opposite beads, these must be of the same colour and the two six other beads must form two mirror images, so there are 6 colourings in each invariant set. For the four reflections whose axes pass between opposite pairs of beads, the four beads on each side must form mirror images, so again there are $\binom{4}{2} = 6$ colourings. Therefore there are $\frac{70+2\times 0+2\times 2+6+4\times 6+4\times 6}{16} = 8$ such necklaces.

3. The board can be treated as being 2-coloured but with a condition on the numbers of colours. As four of the nine squares are filled with Os, there are $\binom{9}{2} = 126$ possible boards.

   The group for rotations is C_4 . If a colouring is invariant under R then there is an X in the centre and the four other Xs either in the corners or at the side midpoints. If a colouring is invariant under R^2 there is X in the centre and 'opposite' squares contain the same symbol. There are four such pairs, two containing X and two O, so there are $\binom{4}{2}$ colourings in the invariant set. Hence $|\Omega_I| = 126, |\Omega_R| = |\Omega_{R^3}| = 2, |\Omega_{R^2}| = 6$ and the number of boards is $\frac{126+2+6+2}{4} = 34$.

   The group with reflections as well is D_4. If a board is invariant under a 'horizontal' reflection, the top and bottom rows are the same. If these rows contain four Os (and the remainder of the board is filled with Xs) there are three possibilities, and if they contain four Xs, then there are three choices for the Os and three for the X in the middle row, so there are nine invariant boards. Hence $|\Omega_X| = |\Omega_Y| = 12$. For the 'diagonal' reflections, a similar argument shows that $|\Omega_U| = |\Omega_V| = 12$. Hence the number of boards is $\frac{126+2+6+2+12+12+12+12}{8} = 23$.

4. (a) The group is C_4 since we divide the grid into four quadrants. As usual, $|\Omega_I| = 2$, the set of all raw colourings. There are 2^4 colourings in the invariant sets for the rotations R or R_3, since each of the four squares in the first quadrant can be coloured in any way. Similarly there are 2^8 colourings in the invariant set for R_2. Hence the number of colourings is $\frac{1}{4}(2^{16} + 2^8 + 2 \times 2^4)$.

   (b) Now the group is D_4. There are 2^8 colourings in the invariant set for reflection in a 'horizontal' or 'vertical' axis, since squares on one side of the axis can be coloured arbitrarily. For a 'diagonal' reflection, the four squares in the diagonal and six squares on one side can be coloured in any way, so the invariant set contains 2^{10} colourings. Hence the number of colourings is $\frac{1}{8}(2^{16} + 2 \times 2^{10} + 3 \times 2^8 + 2 \times 2^4)$.

5. (a) This is the same argument but with more squares and k colours. For rotations, we have $|\Omega_I| = k^{64}, |\Omega_R| = |\Omega_{R^3}| = k^{16}$ and $|\Omega_{R^2}| = k^{32}$. Hence the number of colourings is given by $\frac{1}{4}(k^{64} + k^{32} + 2k^{16})$.

   (b) We now have $|\Omega_X| = |\Omega_Y| = k^{32}$ (for vertical and horizontal axes) and $|\Omega_U| = |\Omega_V| = k^{36}$ (for diagonal axes) so the number of colourings is $\frac{1}{8}(k^{64} + 2k^{36} + 3k^{32} + 2k^{16})$.

6. Turning the flag upside down makes no difference, but we can look at it from either side, so the group consists of the identity and a reflection and is C_2. For the flag with $2n$ stripes, $|\Omega_I| = k^{2n}$ and $|\Omega_R| = k^n$, so there are $\frac{1}{2}k^n(k^n + 1)$ flags. For the flag with $2n + 1$ stripes, $|\Omega_I| = k^{2n+1}$ and $|\Omega_R| = k^n$, so there are $\frac{1}{2}k^n(k^{n+1} + 1)$ flags.

Exercise 3d

1. Label the vertices 1, 2 and 3 anticlockwise. The rotations are $(1\,2\,3)$ and $(1\,3\,2)$, the reflections in axes through 3, 1 and 2 are $(1\,2), (2\,3)$ and $(3\,1)$, and the identity is I. This shows that S_3 and D_3 are effectively the same group.

2. Label the vertices 1, 2, 3 and 4 anticlockwise. The rotations of $90^\circ, 180^\circ$ and 270° are $(1\,2\,3\,4), (1\,3)(2\,4)$ and $(1\,4\,3\,2)$. The reflections in the axes perpendicular to the sides are $(1\,2)(3\,4)$ and $(1\,4)(2\,3)$ and the reflections in the diagonals are $(1\,3)$ and $(2\,4)$.

3. Label the six edges as 12, 13, 14, 23, 24 and 34. It is necessary to do this in order to emphasize which three edges meet at a vertex. The eight rotations of the first type are of the form $(12\,13\,14)(23\,34\,24)$, which is about axis through vertex 1. The three other rotations swap two pairs of edges: for example $(14\,23)(13\,24)$, whose axis is through the midpoints of 12 and 34. The group is the same as before, but now considered as a subgroup of S_6.

4. Such a permutation must consist of disjoint 2-cycles. The number of these is the number of ways of splitting $\{1, 2, \ldots, 2n\}$ into n subsets with two elements in each. This is

$$\frac{\binom{2n}{2} \times \binom{2n-2}{2} \times \cdots \times \binom{4}{2} \times \binom{2}{2}}{n!}$$
$$= (2n-1) \times (2n-3) \times \cdots \times 3 \times 1.$$

5. If the tetrahedron is fixed in space there are three choices per face for which altitude to draw, so we have 81 raw colourings.

   All of these are fixed by the identity.

   None are fixed by a 120° rotation since the axis goes through the centre of a face, so the altitude drawn on that face would move.

   Each of the 180° rotations swaps the positions of two pairs of faces. We may freely choose which altitude to draw on one face in each pair and this completely determines the tetrahedron. Thus there are 9 invariant arrangements for each of these.

   Burnside's lemma now gives a total of $\frac{1\times 81+8\times 0+3\times 9}{12} = 9$ different tetrahedra.

Exercise 3e

1. An octahedron has six vertices, eight faces and twelve edges. A cube, on the other hand, has eight vertices, six faces and twelve edges. The fact that the numbers 6, 8 and 12 appear in both statements suggests that there might be a connection between the two regular polyhedra.

   Mark the centre of each square face of a cube. These six points form the vertices of a regular octahedron as shown on the left in figure 3e.A. Now mark the centre of each triangular face of an octahedron, and join these eight points; the result is a cube as shown on the right in the figure. These facts follow from the symmetry of the situation, or by calculating the distances between adjacent vertices.

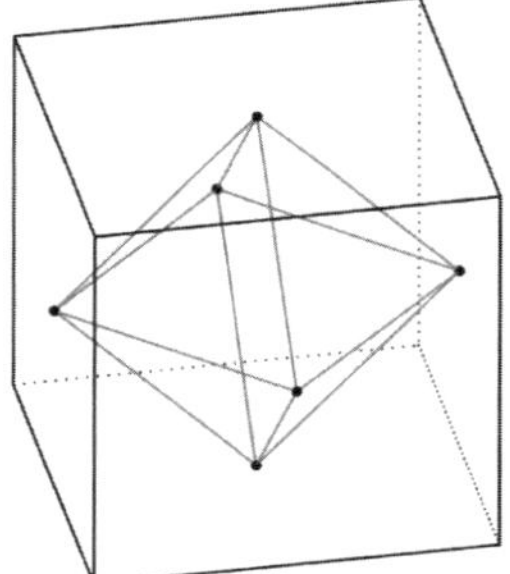

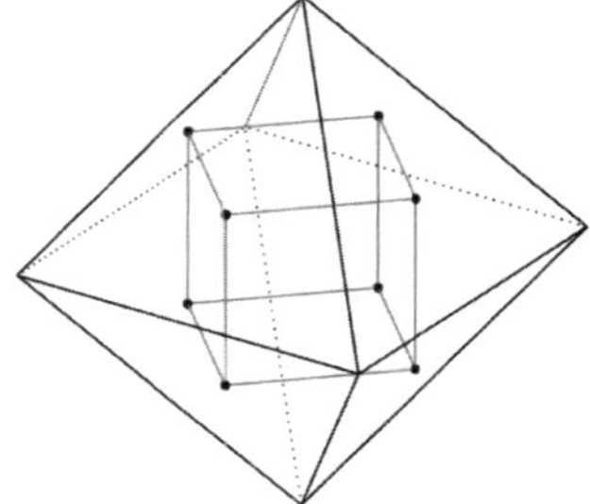

Figure 3e.A

   There is therefore a close relationship between cubes and octahedra: they are said to be *dual* figures. There is a bijection between the two: faces of the one correspond to vertices of the other and edges correspond to edges. It follows that the two solids have the same symmetry group, namely S_4. This also means that transformations which permute the faces of a cube permute the vertices of an octahedron. Hence there is also a correspondence between the cycle indexes. It is straightforward to check that the cube and octahedron have the same edge cycle index, and the vertex cycle index of one solid is the face cycle index of the other.

2. At first sight, it might appear that placing arrows on the edges is equivalent to 2-colouring them. However, when it comes to counting invariant sets, the two things are not equivalent.

In table 3.6, the edge cycle type x_3^4 arises from rotations of type (i), half turns about a long diagonal. Since the three edges 12, 14 and 17 cycle, we have an invariant set if all are coloured red or all are coloured blue. Hence the number of invariant sets arising from this term in the cycle type is 2^4. There is no problem with replacing colours by arrow directions; we could, for instance, regard red as an arrow pointing away from 1 and blue as an arrow pointing towards 1. So here arrows do behave like colours, and the same applies to the cycle types x_4^3 and x_2^6 in cases (ii) and (iii), and also to the identity, when nothing changes.

The difference arises in case (iv) which describes half-turns around an axis through the midpoints of opposite edges. It concerns, in particular, the term which describes the rotation of the edges 14 and 85 into the edges 41 and 58. Colouring an edge 14 red is exactly the same as colouring 41 red; there is no significance in the order of the vertices. But an arrow pointing from 1 to 4 is transformed into one pointing in the opposite direction, and this does matter. It follows there are no invariant sets arising from the edge cycle type $x_1^2x_2^5$. Hence there are 8×2^4 invariant sets for rotations of type (i), 6×2^3 for type (ii), 3×2^6 for type (iii), none for type (iv) and 2^{12} for the identity. By Burnside's lemma, the number of different arrowed cubes is the average size of an invariant set, which is 186.

3. If diagonals are added to the faces, we have a similar issue as in question 2. The problems arise when a face is rotated into itself, indicated by the term x_1^2 in the face cycle types (ii) and (iii). However, we need to look at these separately. In (ii) this corresponds to a face rotation of 90° which would switch the diagonals. Hence the invariant set has no members. In (iii) it is a face rotation of 180° which conserve the diagonal, and so we can count this in the enumeration of invariant sets.

   Hence, by Burnside's lemma, the number of different cubes with face diagonals is $\dfrac{8 \times 2^2 + 6 \times 2^3 + 3 \times 2^4 + 2^6}{24} = 8.$

4. (a) If all the individual dot patterns were unchanged by rotations, then this problem would be equivalent to painting the faces using six different colours. Then the invariant set for any rotation apart from the identity would be empty, so the number of such cubes would be $\frac{6!}{24} = 30$ by Burnside's lemma. However,

the patterns for 2, 3 and 6 dots can each occur in one of two orientations, so this means that the number of such cubes is 240.

(b) We begin by assigning the numbers to each face, and worry about the orientation of the patterns later. Place the cube with the 1 on the bottom and the 6 at the top. Now choose the front face as 2; the rear face is now 5. The 3 and 4 faces can appear in any order, so there are 2 such cubes. Now there are two choices for the orientations of each of the 2, 3 and 6 patterns giving a total of 16 possible cubes.

Exercise 3f

1. Referring to question 3 of exercise 3d, we have the identity of type x_1^4, eight permutations of type $x_1^1x_3^1$ and three of type x_2^2. It follows that the group polynomial in terms of vertices is $P = x_1^4 + 8x_1^1x_3^1 + 3x_2^2$ and the number of k-colourings of vertices is $\frac{1}{12}k^2(k^2+11)$.

2. Again from question 3 of exercise 3d, we have the identity of type x_1^6, eight permutations of type x_3^2 and three of type $x_1^2x_2^2$. Hence the group polynomial in terms of edges is $P = x_1^6 + 8x_3^2 + 3x_1^2x_2^2$ and the number of k-colourings of vertices is $\frac{1}{12}k^3(k^3+3k+8)$.

3. The group is D_8. Number the beads 1 to 8 anticlockwise.
   The identity is of type x_1^8.
   There are four 45° rotations of type x_8^1, two 90° rotations of type x_4^2, a 180° rotation of type x_2^4, four reflections in axes joining opposite pairs of beads of type $x_1^2x_2^3$ and four reflections in axes joining opposite gaps of type x_2^4.
   This yields the number of colourings as $\frac{1}{16}k(k^7+4k^4+5k^3+2k+4)$, which gives the sequence 1, 30, 498, 4435, 25 395 and 107 331 for $k = 1, \ldots, 6$.

4. The group is S_5. It has the vertex cycle index $P_{S_5}^V(x_1, x_2, x_3, x_4, x_5)$ which equals

$$\frac{1}{120}(x_1^5 + 10x_1^3x_1^2 + 20x_1^2x_3^1 + 30x_1^1x_1^4 + 15x_1^1x_2^2 + 20x_2^3 + 24x_5^1).$$

We 'translate' the vertex cycle index into the edge cycle index as in problem 3.12 to produce $P^E_{S_5}(x_1, x_2, x_3, x_4, x_5)$ which equals

$$\frac{1}{120}(x_1^{10} + 10x_1^4x_1^3 + 20x_1^1x_3^3 + 30x_2^1x_4^2 + 15x_1^2x_2^4 + 20x_1^1x_3^1x_6^1 + 24x_5^2)$$

and now we evaluate this by substituting 2 for each variable to obtain 34. This is the number of graphs on five vertices, including the empty graph.

Exercise 3g

1. The group is D_5. Number the vertices 1 to 5 anticlockwise. Then the cycle index is $P = \frac{1}{10}(x_1^5 + 4x_5^1 + 5x_1^1x_2^2)$ which we expand in the form

$$\frac{1}{10}((b+r+g)^5 + 4(b^5+r^5+g^5)^1 + 5(b+r+g)(b^2+r^2+g^2)^2).$$

We have to find the coefficient of b^2r^2g. In the first term, this is $\binom{5}{2\,2\,1}$ which is 30. It does not appear in the second term. In the third, we must choose g in the first bracket and b^2r^2 in the second, and this can be done in two ways. So the coefficient of b^2r^2g in the expression inside the bracket is $30 + 4 \times 0 + 5 \times 2$. Dividing by 10, we see that there are four such pentagons.

2. The group is S_4 and the cycle index is

$$P = \frac{1}{24}(x_1^6 + 6x_1^2x_4^1 + 3x_1^2x_2^2 + 8x_3^2 + 6x_2^3)$$

so we have to find the coefficient of $b^2r^2g^2$ in the expression

$$(b+r+g)^6 + 6(b+r+g)^2(b^4+r^4+g^4) + \cdots + 6(b^2+r^2+g^2)^3.$$

This is $\binom{6}{2\,2\,2} + 0 + 3 \times 3 \times 2 + 0 + 6 \times 6 = 144$ so after dividing by 24 we see that there are 6 such cubes.

3. Now the cycle index is $P = \frac{1}{24}(x_1^{12} + 6x_4^3 + 8x_3^4 + 3x_2^6 + 6x_1^2x_2^5)$ and we have to find the coefficient of $b^4r^4g^4$ in the expression

$$(b+r+g)^{12} + 6(b^4+r^4+g^4)^3 + \cdots + 6(b+r+g)^2(b^2+r^2+g^2)^5$$

which is $\binom{12}{4\,4\,4} + 6 \times 6 + 0 + 3 \times \binom{6}{2\,2\,2} + 6 \times 3 \times \binom{5}{1\,2\,2} = 35\,496$. Dividing by 24, we obtain 1479 such cubes.

4. The cycle index from question 4 of exercise 3f is modified to obtain

$$P^E_{S_5}(x) =$$
$$(x+1)^{10} + 10(x+1)^4(x^2+1) + 20(x+1)(x^3+1)^3$$
$$+ 30(x^2+1)(x^4+1)^2 + 15(x+1)^2(x^2+1)^4$$
$$+ 20(x+1)(x^3+1)(x^6+1) + 24(x^5+1)^2$$

which expands to give

$$P^E_{S_5}(x) = x^{10} + x^9 + 2x^8 + 4x^7 + 6x^6 + 6x^5 + 6x^4 + 4x^3 + 2x^2 + x + 1.$$

This tells us that there is one graph of order five with ten edges, one with nine, two with eight, four with seven, six with six, six with five, six with four, four with three, two with two, one with one and one (the empty graph) with none. It is encouraging to discover that this gives a total of 34 graphs.

Exercise 4a

1. Take the Ferrers diagram for the partition of $n+k$ into k parts and remove the first column, which has exactly k rows. The result is a partition of n into at most k parts. This is a bijection, since an appropriate column can always be added to the left of a Ferrers diagram for a partition of n into at most k parts, creating additional rows as necessary at the bottom.

2. Take the Ferrers diagram for a partition of n into at most k parts. Add k dots to the top row, $k-1$ to the second, and so on down to 1 dot, creating additional rows as necessary. The result is a partition of $n+\frac{1}{2}k(k+1)$ into k parts. These must be distinct, since if not then the rows in the original diagram would not have been in decreasing order of length. This process can be reversed.

   Alternatively, use the result of question 1 of this exercise. Start with the Ferrers diagram for a partition of $n+k$ into k parts and add $k-1$ dots to the top row, $k-2$ to the second, and so on.

3. Observe that $p(n) - p^k(n)$ is the number of partitions of n which do use k. It is clear that these can all be obtained from the partitions of $n-k$ by adding a part k, so there is a bijection between these and the partitions of $n-k$. Note that this is true for $k=n$ as long as we define $p(0)=1$.

   Alternatively, note that the generating function for $p^k(n)$ is given by $(1-x^k)E(x)$ (where we ignore terms where $k > n$). The coefficient of x^n in this series is obtained by subtracting the coefficient of x^{n-k} from that of x^n, so the result follows.

4. We use generating functions. The coefficient of x^n in $G_1(x)$ is the number of partitions of n in which no part is repeated more than k times, and that in $G_2(x)$ is the number of partitions of n with no

part being a multiple of $k+1$.

$$\begin{aligned}
&G_1(x)\\
&= (1+x+\dots+x^k)(1+x^2+\dots+x^{2k})\dots(1+x^r+\dots+x^{rk})\dots\\
&= \prod_i (1+x^i+\dots+x^{ik})\\
&= \prod_i \frac{1-x^{i(k+1)}}{1-x^i}\\
&= \prod_{i\neq j(k+1)} \frac{1}{1-x^i}\\
&= \prod_{i \text{ not a multiple of } k+1} (1+x^i+x^{2i}+\dots)\\
&= G_2(x).
\end{aligned}$$

5. Let u_r be the rth pentagonal number.

   We use induction, with the hypothesis that $u_k = \frac{1}{2}k(3k-1)$. Clearly $u_1 = 1$ satisfies this formula. When we move from u_k to u_{k+1}, we superimpose a new pentagon with $k+1$ dots on each side. This contains $5k$ dots, but $2k-1$ of them are already there, so there are $3k+1$ new dots. It follows that

   $$u_{k+1} = \frac{1}{2}k(3k-1) + 3k + 1 = \frac{1}{2}(k+1)(3(k+1)-1)$$

   and the result follows by induction.

6. Let v_r be the number of dots inside the pentagon with side $r+2$.

   $$v_r = u_{r+2} - 5(r+1) = \frac{1}{2}(r+2)(3r+5) - 5(r+1) = \frac{1}{2}r(3r+1).$$

7. The zero term of the sum gives the value 1. When $n = -m$, the term is $(-1)^{-m}x^{-\frac{1}{2}m(-3m-1)} = (-1)^m x^{\frac{1}{2}m(3m+1)}$, producing the other pentagonal numbers as indices. Hence the sum can be written as $1 - x - x^2 + x^5 + x^7 - \dots$ as required.

Exercise 4b

1. (a) Place the element 1 in one of the two sets, and now place subsequent elements from 2 to n in either set. This gives 2^{n-1}

partitions, but it includes one in which every element is in the same set, which we must subtract.

(b) Place the elements from 1 to n in one of the three sets, yielding 3^n choices. Using PIE, we exclude the 3×2^n choices where at most two sets are used, and include the 3 choices where only one set has been used. However, every possibility has been counted 3! times. This yields the required formula.

(c) A partition into $n-1$ sets consists of one set of 2 elements and the rest are singleton sets. Hence the number of partitions is $\binom{n}{2}$.

(d) This does not actually say anything original; it simply describes the problem. The multinomial coefficient is the number of ways of partitioning the set $\{1, 2, \ldots, n\}$ into an ordered sequence of k disjoint subsets with sizes $n_1, n_2, \ldots, n_k$, and we want the number where the order is irrelevant, so we should divide by $k!$ to avoid overcounting.

2. Consider a partition of $\{1, 2, \ldots, n\}$ into k disjoint subsets. The element 1 will be in one of these, together with r other elements, where $0 \leq r \leq n-1$. They can be chosen in $\binom{n-1}{r}$ ways. The other $n-r-1$ elements are partitioned into $k-1$ disjoint subsets, and this can be done in $S(n-r-1, k-1)$ ways. Hence, for this value of r, there are $\binom{n-1}{r} S(n-r-1, k-1)$ partitions into k sets, and the total number of such partitions is found by summing over r between 0 and $k-1$.

3. By the result of question 1 (d) of this exercise, this is equivalent to $\sum_{r=1}^{n} \binom{N}{r} \left(\sum \binom{n}{n_1\, n_2 \cdots n_r} \right)$, where the inner sum ranges over all the compositions of n into r parts. This is equivalent to considering a row of n boxes, into which we put r different numbers in the range 1 to N. The boxes chosen (in order) are counted by the multinomial coefficient, and the numbers used by the binomial coefficient. If, for example, we decide to use only 1s , 2s and 3s, this is the result of a choice of three numbers from N, together with a specification of which boxes they should occupy (in order). The full set of choices

(after the summations have been carried out) is simply the number of ways in which N numbers can be placed in n boxes, which is N^n. Alternatively, we could use induction on n, taking an arbitrary $N \geq n$. The base case is trivial, so suppose that the statement is proved for $n = k$. Now the general term of the summation for $n = k + 1$ is

$$r!\binom{N}{r}S(k+1,r) = r!\binom{N}{r}(S(k,r-1) + rS(k,r))$$

by the recurrence relation (4.3). We regroup the terms to produce a general summand

$$\left((r+1)!\binom{N}{r+1} + r \times r!\binom{N}{r}\right)S(k,r)$$

and this simplifies to

$$\begin{aligned}\left((r+1)!\binom{N}{r+1} + r \times r!\binom{N}{r}\right)S(k,r) &= [(N-r)+r]r!\binom{N}{r}S(k,r)\\ &= Nr!\binom{N}{r}S(k,r)\end{aligned}$$

and so

$$\sum_{r=1}^{k+1} r!\binom{N}{r}S(k+1,r) = N \times \sum_{r=1}^{k} r!\binom{N}{r}S(k,r) = N^{n+1}$$

by the induction hypothesis. The result follows.

It would be nice to find a purely combinatorial proof of this.

4. We begin by populating the first few rows of the array.

$$\begin{array}{llll} a_0 & & & \\ a_1 & a_0 + a_1 & & \\ a_2 & a_1 + a_2 & a_0 + 2a_1 + a_2 & \\ a_3 & a_2 + a_3 & a_1 + 2a_2 + a_3 & a_0 + 3a_1 + 3a_2 + a_3 \end{array}$$

   (a) We see that $u(n,r) = 2^r$. This is easily proved by induction.

   (b) It is obvious what is going on here. The coefficients in the rth column of the nth row are the entries in the rth row of Pascal's triangle, and these are attached to terms of the sequence $a_0\, a_1\, a_2\, \ldots\, a_n$ beginning at a_{n-r}. The formula is therefore

$$u(n,r) = \sum_{i=0}^{r} \binom{r}{i} a_{n-r+i} \tag{6.1}$$

which can be seen to do exactly what we require. Again the proof is a rather awkward induction. The step consists of a calculation akin to $(a_1 + 2a_2 + a_3) + (a_2 + 2a_3 + a_4)$, and it is clear that the induction hypothesis follows from Pascal's identity. Writing this out in general terms would be a technical exercise rather than a matter of enlightenment.

(c) We may use equation (6.1) from (b) to obtain $u(n,n) = B_{n+1}$.

(d) We define $a_0 = 1$ and $a_{n+1} = u(n,n) = B_{n+1}$, so the array is allowed to 'seed' itself and produce Bell numbers as the final entries in each column. Equivalently, we can forget about the a_i and define the array by

$$\begin{aligned} u(0,0) &= 1 \\ u(n+1,0) &= u(n,n) \qquad \text{for } n \geq 0 \\ u(n+1,r+1) &= u(n,r) + u(n+1,r) \qquad \text{for } n \geq 0, 0 \leq r \leq n \end{aligned}$$

and the result will be that $u(n,n) = u(n+1,0) = B_{n+1}$.

(e) Define $v(n,r)$ (for $n \geq 0, 0 \leq r \leq n$) as the number of partitions of $\{1,2,\ldots,n+2\}$ including $\{r+2\}$ as the 'largest' singleton. We shall show that $v(n,r) = u(n,r)$ as in (d). We will call a partition with no singletons *singleton-free* (SF).
First we consider $v(0,0)$, which is the number of partitions of $\{1,2\}$ with largest singleton $\{2\}$. There is exactly one of these, namely $\{1\} \cup \{2\}$, so $v(0,0) = u(0,0)$.
Now we consider $v(n,0)$ for $n > 0$, which is the number of partitions of $\{1,2,\ldots,n+2\}$ with largest singleton $\{2\}$. We have to show that this is B_n. This is the hardest part of the proof, and it might clarify matters for the reader if they work through a particular case with, say, $n = 4$. What we are doing is showing that the number of partitions of $\{1,2,\ldots,n\}$ is

the sum of the numbers of SF partitions of $\{1, 2, \ldots, n\}$ and $\{1, 2, \ldots, n+1\}$.
If a partition of $\{1, 2, \ldots, n+2\}$ with largest singleton $\{2\}$ contains $\{1\}$, we can obtain a SF partition of $\{3, 4, \ldots, n+2\}$ by omitting the $\{1\}$. Now we reduce each element of this by 2 to produce a SF partition of $\{1, 2, \ldots, n\}$.
If it does not contain $\{1\}$, then we have a SF partition of $\{1, 3, 4, \ldots, n+2\}$. In this case, we first reduce each element greater than 1 by 2, and then replace each part which includes $n+1$ by the union of singletons for the other elements in it, but leaving out $\{n+1\}$. We thus obtain a partition of $\{1, 2, \ldots, n\}$ which is not SF.
Together, these partitions are all the partitions of $\{1, 2, \ldots, n\}$; this can be seen by reversing the steps. Now we have shown that $v(n, 0) = B_n$.
Next we show the recursion step. Note that $v(n+1, r+1)$ is the number of partitions of $\{1, 2, \ldots, n+3\}$ with largest singleton $\{r+3\}$. If $\{r+2\}$ is a singleton, we can delete the element $r+3$ and subtract 1 from each $k > r+3$, and the result is a partition of $\{1, 2, \ldots, n+2\}$ with largest singleton $\{r+2\}$. There are $v(n, r)$ of these. If, on the other hand, $\{r+2\}$ is not a singleton, we can simply swap the elements $\{r+2\}$ and $\{r+3\}$, and now we have a partition of $\{1, 2, \ldots, n+3\}$ with largest singleton $\{r+2\}$. There are $v(n+1, r)$ of these. Therefore $v(n+1, r+1) = v(n, r) + v(n+1, r)$.
This establishes that $u(n, r)$ has the required property.

5. Without loss of generality, we can take N to be $\prod_{r=1} p_r$, the product of the first n prime numbers. Then each representation of N as a product of integers greater than one corresponds to a partition of $\{2, 3, 5, \ldots, p_n\}$ into disjoint non-empty subsets, and hence the number of such products is B_n.

6. Consider the set $\{1, 2, 3, \ldots, n\}$ which represents the lines of the poem. Now a rhyme-scheme assigns letters to these numbers, and this can be done by partitioning the set into disjoint non- empty subsets, placing them in order of the smallest element and then giving these subsets letter names in alphabetical order. Hence the number of rhyme-schemes is B_n.

For example, the first example would correspond to the partition $\{1,2,5\},\{3,4\}$ and the second to $\{1\},\{2\},\{3\},\{4\},\{5\}$.

Exercise 4c

1. (a) The permutation is an n-cycle and there are $(n-1)!$ of these.

   (b) There are $(n-2)$ 1-cycles and one 2-cycle, which can be chosen in $\binom{n}{2}$ ways.

   (c) There are $(n-3)$ 1-cycles and one 3-cycle, which can be chosen in $2\binom{n}{3}$ ways, or there are $(n-4)$ 1-cycles and two 2-cycles, which can be chosen in $3\binom{n}{4}$ ways. Hence the total number of such permutations is $\binom{n}{3}[2+\frac{3}{4}(n-3)] = \frac{1}{4}(3n-1)\binom{n}{3}$.

   (d) The alternatives are:
   $(n-4)$ 1-cycles and one 4-cycle: $6\binom{n}{4}$ ways;
   $(n-5)$ 1-cycles, one 2-cycle and one 3-cycle: $10 \times 2\binom{n}{5}$ ways;
   $(n-6)$ 1-cycles, three 2-cycles: $5 \times 3\binom{n}{6}$ ways.
   Adding and rearranging we obtain $\binom{n}{2} \times \binom{n}{4}$.

   (e) Using a similar analysis, we obtain the equation
   $S(n,2) = \frac{1}{2}\sum(n-r-1)!(r-1)!\binom{n}{r}$.
   This reduces to $\frac{1}{2}(n-1)!\sum_{r=1}^{n-1}\left(\frac{1}{r}+\frac{1}{n-r}\right)$ which, after some rearrangement, is the desired expression.

2. Use the 'coding' method for Stirling numbers of the first kind in the text. You have a row of n cells of which k have been selected. The number of choices for each cell between A and B is 1, that for each cell between B and C is 2, and so on, with the number of choices for each cell at the right-hand end being k. It is now clear that the generating function $G_k(x) = \sum S(n,k)x^n$ is given by the product

$$x^k(1+x+x^2+\dots)(1+2x+4x^2+\dots)\dots(1+kx+k^2x^2+\dots)$$

   which can be rewritten in the desired form using the sum to infinity of a GP.

   It is quite feasible to use this to calculate the value of $S(6,3)$, for example.

3. See table 4c.A.

	D boxes D balls	I boxes D balls	D boxes I balls	I boxes I balls
Unrestricted	B	I	E	K
At most one ball per box	C	A	D	A
At least one ball per box	H	G	F	J

Table 4c.A

Exercise 5a

1. This game is similar to problem 5.1 but is a little more subtle. The game graph is shown in figure 5a.A and shows that 21 counters is an $\mathcal{L}$-position so Bob can win. It also appears that the sequence of $\mathcal{W}$ and $\mathcal{L}$-positions is periodic with period seven. This can be verified by a routine induction argument.

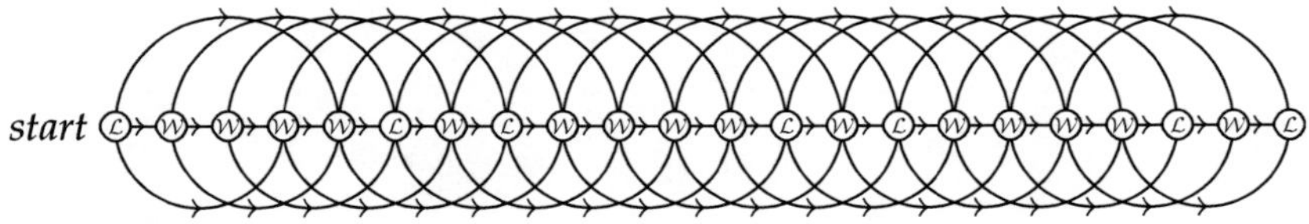

Figure 5a.A

2. This is something of a trick question. Each move changes the parity of the number of counters on the table, so zero can only be reached after an odd number of moves. Therefore Alice is sure to win, regardless of how either she or Bob chooses to play.

3. Rather than draw the full game graph, we simply label the points as shown in figure 5a.B. The starting position is an $\mathcal{L}$-position so Bob has a winning strategy. Bob simply moves the counter onto the leading diagonal of the grid every turn. (Alice will have to move the counter off the diagonal on each of her turns.)

$\mathcal{W}_2$	$\mathcal{W}_4$	$\mathcal{W}_6$	$\mathcal{W}_8$	$\mathcal{W}_{10}$	$\mathcal{W}_{12}$	$\mathcal{W}_{14}$	$\mathcal{L}_{15}$
$\mathcal{W}_2$	$\mathcal{W}_4$	$\mathcal{W}_6$	$\mathcal{W}_8$	$\mathcal{W}_{10}$	$\mathcal{W}_{12}$	$\mathcal{L}_{13}$	$\mathcal{W}_{14}$
$\mathcal{W}_2$	$\mathcal{W}_4$	$\mathcal{W}_6$	$\mathcal{W}_8$	$\mathcal{W}_{10}$	$\mathcal{L}_{11}$	$\mathcal{W}_{12}$	$\mathcal{W}_{12}$
$\mathcal{W}_2$	$\mathcal{W}_4$	$\mathcal{W}_6$	$\mathcal{W}_8$	$\mathcal{L}_9$	$\mathcal{W}_{10}$	$\mathcal{W}_{10}$	$\mathcal{W}_{10}$
$\mathcal{W}_2$	$\mathcal{W}_4$	$\mathcal{W}_6$	$\mathcal{L}_7$	$\mathcal{W}_8$	$\mathcal{W}_8$	$\mathcal{W}_8$	$\mathcal{W}_8$
$\mathcal{W}_2$	$\mathcal{W}_4$	$\mathcal{L}_5$	$\mathcal{W}_6$	$\mathcal{W}_6$	$\mathcal{W}_6$	$\mathcal{W}_6$	$\mathcal{W}_6$
$\mathcal{W}_2$	$\mathcal{L}_3$	$\mathcal{W}_4$	$\mathcal{W}_4$	$\mathcal{W}_4$	$\mathcal{W}_4$	$\mathcal{W}_4$	$\mathcal{W}_4$
$\mathcal{L}_1$	$\mathcal{W}_2$	$\mathcal{W}_2$	$\mathcal{W}_2$	$\mathcal{W}_2$	$\mathcal{W}_2$	$\mathcal{W}_2$	$\mathcal{W}_2$

Figure 5a.B

4. We start by a smaller version of the same problem. The labelled game graph for up to ten counters is shown in figure 5a.C.

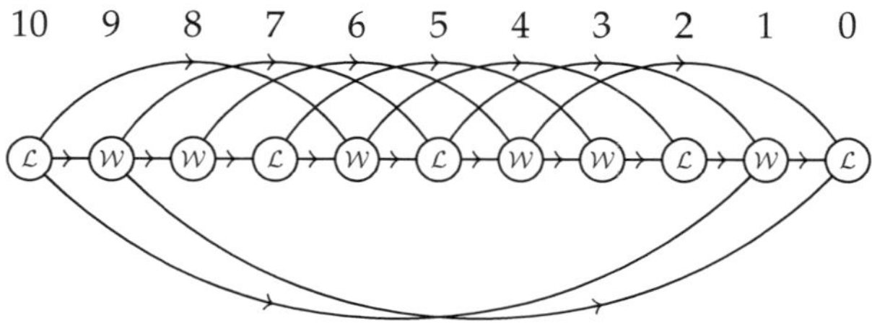

Figure 5a.C

This shows that ten counters is an $\mathcal{L}$-position.

This means that Alice can win by taking one hundred counters on her first turn.

Exercise 5b

1. There are a huge number of possible opening moves for Alice (2 014 025 in fact), so drawing a game graph is out of the question. Alice might try a move which limits the number of options Bob has on his turn. She might, for example, write 2^{2007} and 5^{2007} on the board since then Bob has no choice but to split a prime power into two smaller powers of that prime. At this point Alice might notice that there is a symmetry between powers of two and powers of five which she may be able to exploit. If Bob splits p^{2007} into p^a and p^b, Alice can respond by splitting q^{2007} into q^a and q^b where $\{p, q\} = \{2, 5\}$. In fact, Alice can always mimic Bob's moves in this way, so Bob can never make the last move in the game.

2. Bob's first thought might be to copy each of Alice's moves so as to ensure he makes the last one. For example, he might try to preserve one of the symmetries of the original rectangular board. This is a good idea, but does not provide Bob with a move if Alice places a domino on the middle two cells of the board.

   In fact, playing in the centre of the board is a winning first move for Alice since she can now use a symmetry strategy to defeat Bob. Wherever he plays she can place a domino so as to ensure that the position after her turn always has rotational symmetry of order two. This means Alice is sure to make the last move of the game.

This argument is completely general provided the starting board has two central cells for Alice's first move. This is true for any rectangle with one odd side length and one even side length.

3. Now the strategy that Bob tried on the 3×4 grid works. He can use each of his turns to ensure the position retains rotational symmetry of order two. This ensures he will make the last move in the game.

   This argument works for any rectangular starting grid where both of the dimensions are even.

4. Since the question involves factors and multiples, it is natural to rewrite 160 000 as $2^8 5^4$. Now all the factors are of the form $2^a 5^b$ and there are $(8+1)(4+1) = 45$ such factors. If a player writes $2^{x_0} 5^{y_0}$ on the board then $2^x 5^y$ becomes illegal for all $x \geq x_0$, $y \geq y_0$. Now we see that the game is actually entirely equivalent to chomp on a 9×5 board so problem 5.8 shows that Alice has a winning strategy.

   The game of chomp was actually first studied in this form and only reinterpreted as a game about chocolate bars much later.

5. (a) Initially every row and column on the chocolate bar contains infinitely many cells. However if Alice's first move is to *chomp* (x_0, y_0), that is, to remove cells centred at (x, y) for $x \geq x_0$, $y \geq y_0$, then only $x_0 - 1$ columns and $y_0 - 1$ rows are still infinite (meaning they contain infinitely many cells which have not yet been removed). We focus our attention on I, the total number of rows and columns which are infinite. After Alice's first turn $I = x_0 + y_0 - 2$ which is finite. Bob must now choose a cell which belongs to an infinite row or column, so his move will decrease I. It would be nice if I decreased every turn, since then it would reach zero after at most $x_0 + y_0 - 2$ turns, and once $I = 0$ the number of moves left in the game is bounded above by the number of cells which remain.

   Unfortunately, Alice's second turn might not decrease I, since she might be able to chomp a cell which does not belong to an infinite row or column. However, there are only ever finitely many cells which do not belong to an infinite row or column. Thus, while I may not decrease every turn, it can never increase and cannot be constant for an infinite number of turns. This means the game is sure to terminate after finitely many turns.

We note that although the game always ends there is no integer N with the property that the game ends after at most N turns.

(b) If the game had admitted infinite sequences of moves, then it might have been the case that both players could ensure they never lost. As it is one player is sure to have a winning strategy, but the strategy stealing argument used in problem 5.8 no longer works because there is no move analogous to removing only one cell.

Let us first see whether Alice has any sensible opening moves. She has two possible guiding principles: one is to limit Bob's options and the other is to create symmetry she can later exploit. Either (or both) of these might lead her to chomp $(2,2)$ leaving only the infinite L-shape formed by (x,y) where $\min(x,y) = 1$. This turns out to be an inspired move. Bob's move will be in one arm of the L and Alice can mirror it in the other. This symmetry strategy ensures Alice will win.

6. We may assume, without loss of generality, that Alice's first number is at most $\frac{n}{2}$. (If she chooses a larger number, relabel each number x with $n+1-x$.)

Bob can choose n on his first turn. From then on, he can never be left without a move, since he can always choose a number between two numbers Alice has chosen. If he happens to choose a number smaller than Alice's smallest number, or larger than her largest, this only helps him. Therefore, Bob never loses the game.

It remains to check when Alice can force a draw.

The game is certainly a draw for $n = 1$, but for other odd numbers Bob is sure to win, since a draw would mean Alice made the last move which never happens.

For $n = 2$ the game is always a draw, and for $n = 4$ Alice can force a draw by choosing 1 on her first turn.

For $n = 6$ Alice can also force a draw. She starts with 1. If Bob does not choose 6 then she can choose it on her second turn. On her third turn she can choose a number between Bob's numbers which ensures she will not lose. (In fact, Bob will lose in this case.)

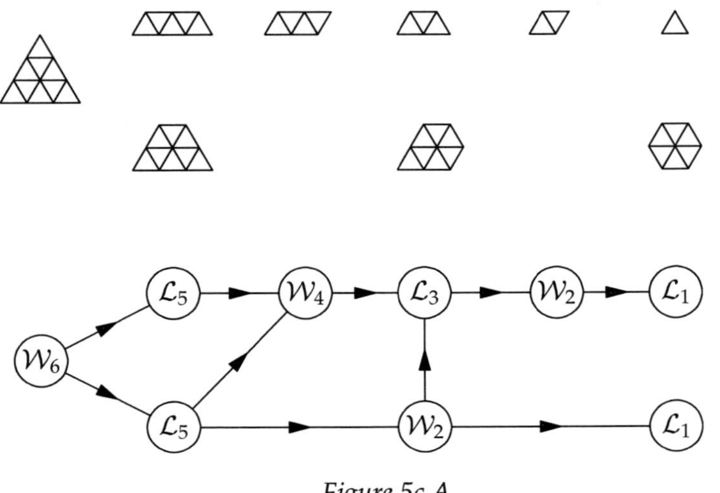

Figure 5c.A

If, on the other hand, Bob does choose 6, then Alice can get a draw by choosing 3 on her second turn and 5 on her third.

Finally we claim that Bob can force a win for all even numbers $n > 6$. On his first turn he chooses n. Now Alice can only get a draw if she manages to choose all the odd numbers. However, by the end of her second turn, Alice can have at most two of the numbers 1, 3, 5 so Bob can choose one of these on his second turn.

Exercise 5c

1. For $n = 2$ the game always last three turns so Alice wins. The game graph for $n = 3$ is shown in figure 5c.A. We see that the starting position is a $\mathcal{W}$-position, so Alice has a winning strategy.

   More importantly her strategy is easy to describe. She takes a piece of side length two leaving a row of five small triangles. She and Bob now eat these one at a time with Bob going first. Since five is an odd number Alice will eat the penultimate piece and win.

   This strategy works for any value of n. Alice starts by eating the largest piece she can. This leaves a row of $2n - 1$ small triangles, which will be eaten one at a time until one is left after Alice's turn.

We note that Alice's strategy hinges on making the opening move which most severely limits Bob's options.

2. For $n = 3$ the game is certain to last two turns, so Bob wins. For $n = 4$ Bob can win on his first turn by moving to two piles of two nuts. For $n = 5$ the game is sure to last four turns so Bob wins.

   For $n = 6$ Bob can either move to $(2, 2, 1, 1)$ or $(3, 1, 1, 1)$ on his first turn, but the first option is fatal, since Alice will move to $(2, 2, 2)$. If Alice is facing $(3, 1, 1, 1)$ she can move to $(3, 2, 1)$ or $(4, 1, 1)$. In the former case Bob can move to $(4, 2)$ or $(3, 3)$ and win, and in the latter he can move to $(4, 2)$ and win. (He could also move to $(5, 1)$ in either case and lose.)

   For $n = 7$ the game is sure to last six turns, since the only position where no two piles have a coprime number of nuts is a single pile. Thus Bob is sure to win.

   For $n = 8$ it is not too hard to draw a full game graph, and discover that, yet again, Bob can win. However, we instead follow the advice given before the exercise and consider positions with few followers. A position has few followers if many piles share a factor and cannot be combined, or if many piles have the same number of nuts so are equivalent. Since the initial position has many piles of one nut, there may be mileage in a strategy which keeps as many piles of one nut as possible.

   If Bob follows this strategy he will move to $(3, 1, 1, 1, 1, 1)$ on his first turn. Alice has two options, but either way Bob can move to $(5, 1, 1, 1)$. Again Alice has two options namely $(6, 1, 1)$ and $(5, 2, 1)$. From here Bob could go to $(7, 1)$ and lose, but he can also go to $(6, 2)$ and win.

   For $n = 9$ Bob can again follow this strategy. If he does then the positions after his turns will be $(3, 1, 1, 1, 1, 1, 1)$, then $(5, 1, 1, 1, 1)$, then $(7, 1, 1)$ and finally (9).

   Now we have enough for a general strategy. If n is odd, Bob can win with exactly the same strategy he used for $n = 9$ and ensuring the maximum possible number of piles contain a single nut. If, on the other hand, n is even, then Bob starts in the same way until he reaches the position $(n - 3, 1, 1, 1)$. Alice now goes to either $(n - 2, 1, 1)$ or $(n - 3, 2, 1)$ and Bob wins by going to $(n - 2, 2)$.

3. For $n = 2$, the smallest number of stones a player cannot take in one turn is nine. Zero stones is an $\mathcal{L}$-position, so $s = 1, 2, 3, \ldots, 8$ are all $\mathcal{W}$-positions. This means $s = 9$ is an $\mathcal{L}$-position. All greater values of s are $\mathcal{W}$-positions since it is always possible to move to either $s = 0$ or $s = 9$ by removing an even number of stones.

   For $n = 3$ we have that $s = 0$ is $\mathcal{L}$ and $s = 1, 2, 3$ are all $\mathcal{W}$. This means $s = 4$ is $\mathcal{L}$. The positions $s = 5, 6, 7$ all have $s = 4$ among their followers, so they are all $\mathcal{W}$ which implies that $s = 8$ is $\mathcal{L}$. Now we have three $\mathcal{L}$-positions, $s = 0, 4, 8$. If $s > 8$ it is possible to reach one of these positions by removing a multiple of three stones, since these three values of s represent all three residue classes modulo 3. Thus there are two values of s where Bob has a winning strategy.

   For $n = 4$ we proceed in a similar way. The first four $\mathcal{L}$-positions we encounter are $s = 0, 6, 15, 21$ and these values of s represent all four residue classes modulo four. Therefore, Bob can win precisely when $s = 6, 15$ or 21 initially. For any larger value of s Alice can win by leaving either 0, 6, 15 or 21 stones at the start of Bob's turn.

   By now we suspect that there are always $n - 1$ values of s where Bob can win the game, one for each residue class modulo n apart from 0. Certainly if $s = a$ is an $\mathcal{L}$-position, then $s = a + kn$ is a $\mathcal{W}$-position for all $k > 0$. Thus there are at most n different $\mathcal{L}$-positions in the game since no two are congruent modulo n.

   Let us call the $\mathcal{L}$-positions $s_1, s_2, \ldots$ and suppose, for contradiction, that there are fewer than n of them. This implies that there is some congruence class modulo n which does not contain an $\mathcal{L}$-position. Since every member of this congruence class is a $\mathcal{W}$-position, each one must be of the form $s_i + p$ where s_i is an $\mathcal{L}$-position and p is prime. To put this another way, for any N in this congruence class at least one of the numbers $N - s_1, N - s_2, \ldots$ must be prime.

   This is implausible since we know the primes become progressively rarer as we move up the number line. In particular it is well known that for any $T > 1$ the numbers $T! + 2, T! + 3, \ldots, T! + T$ are all composite, so there exist arbitrarily long sequences of composite numbers.

   We can use this fact to complete the proof. We choose T to be larger than $n + s_i$ for any i and find a block of $T - 1$ consecutive composite numbers. Now we choose N to be the largest number in this block such that none of the s_i are congruent to N modulo n. Our choice of T means that $N - s_i$ is composite for all i which is precisely the contradiction we had hoped to reach.

4. First we must determine which numbers can be expressed as the difference of two squares. The equations $2n + 1 = (n + 1)^2 - n^2$ and $4n = (n + 1)^2 - (n - 1)^2$ show that all odd numbers and multiples of four can be written as the difference of two squares. On the other hand, if $a^2 - b^2 = (a - b)(a + b)$ is even, then one and therefore both of the expressions $(a - b)$ and $(a + b)$ must be even, so $a^2 - b^2$ is a multiple of four. Thus the integers not expressible as the difference of two squares are precisely those congruent to 2 modulo 4.

   This means we are free to consider the integers from 1 to 100 modulo 4 and may imagine Naomi and Tom playing the same game with twenty-five copies of each of the numbers 0, 1, 2 and 3.

   With this in place we can describe a winning strategy for Naomi. She begins by choosing a 0. This means Tom may not choose a 2. If he chooses a 0, then Naomi does likewise, and if he chooses a 1 she chooses a 3 and vice versa. This strategy ensures that at the end of each of Naomi's turns the total of all numbers chosen is zero modulo 4.

   It is also the case that after Naomi's turn the number of 0s remaining is even, and the numbers of 1s and 3s remaining are equal. This means that when no more 0s, 1s or 3s are left, it is Tom's turn. He must now choose a 2 and lose.

5. The game ends precisely when both a and b are odd, so such positions are a win for Bob. We also note that if $a + b$ is odd, the game is sure to last for ever. From now on we assume that a and b are even initially.

   Our next observation is that if a or b is congruent to 2 modulo 4 then Alice can immediately move to a position where both piles are odd and win.

   We are left with the case where both a and b are divisible by 4 and may write $a = 4a'$ and $b = 4b'$. Now if a' and b' are both odd, then after Alice's first turn the numbers of coins in the piles will be 2 modulo 4, at which point Bob can win.

   If exactly one of a' and b' is odd, then we may assume it is a'. Halving the $4a'$ pile is fatal for Alice, so after her turn the position will be $(4(a' + \frac{b'}{2}), 4(\frac{b'}{2}))$. Both numbers are still divisible by 4, and since the total has not changed, exactly one must be 4 times an odd number. Thus we are in exactly the same sort of position as we were before, but with Bob to play. He must behave as Alice did to avoid losing immediately and we will end up in an infinite loop.

We are left to consider the position $(4a', 4b')$ where a' and b' are both even. If either of a' or b' is congruent to 2 modulo 4, then Alice may use her first turn to move to a position where both piles contain four times an odd number of coins. Our earlier work shows this is a win for Alice. (Bob will move to a position where both numbers are 2 modulo 4 and then lose.)

So far it seems as though the position $(4a', 4b')$ has the same outcome as (a', b'), and this suggests the following outcome profile.

Consider $v_2(a)$ and $v_2(b)$.

If these numbers are equal and even, Bob wins, while if these numbers are equal and odd, Alice wins.

If $\min(v_2(a), v_2(b))$ is even (and $v_2(a) \neq v_2(b)$), the game lasts forever.

If $\min(v_2(a), v_2(b))$ is odd, Alice wins. (This also covers the case when $v_2(a) = v_2(b)$.)

Once we have articulated this conjecture, it is fairly straightforward to verify it.

If $v_2(a) = v_2(b)$ and we start at $(2^c x, 2^c y)$ where $c \geq 1$ and x and y are odd, then, without loss of generality, the next position is $(2^{c-1}x, 2^{c-1}(2y + x))$ so the values of $v_2(a)$ and $v_2(b)$ remain equal and decrease by one each turn which establishes the first part of the outcome profile. This could be made more formal by a routine induction.

Next suppose $\min(v_2(a), v_2(b)) = 2c + 1$ so the initial position is $(2^{2c+1}x, 2^{2c+1}y)$ where x is odd and y may or may not be.

Alice can move to $(2^{2c}x, 2^{2c}(x + 2y))$ and she is sure to win from there.

Finally we assume the position is $(2^{2c}x, 2^{2c+1}y)$ where x is odd and y may or may not be. Moving to $(2^{2c-1}x, 2^{2c-1}(x + 4y))$ would be fatal for Alice so she will move to $(2^{2c}(x + y), 2^{2c}y)$. We know x is odd, so y and $x + y$ have opposite parity which implies we are in the same type of position as we were initially. Bob will now act as Alice did and the game will enter an infinite loop.

Thus (a, b) is a win for Bob precisely when $v_2(a) = v_2(b) = 2k$ for some integer k.

Exercise 5d

1. (a) $[15, 16, 17]$ or

$$\begin{aligned} 15 &=_2 01111 \\ 16 &=_2 10000 \\ 17 &=_2 10001 \end{aligned}$$

is a $\mathcal{W}$-position. The only $\mathcal{L}$-position which follows it is

$$\begin{aligned} 1 &=_2 00001 \\ 16 &=_2 10000 \\ 17 &=_2 10001 \end{aligned}$$

or $[1, 16, 17]$.

(b) $[5, 10, 15]$ or

$$\begin{aligned} 5 &=_2 0101 \\ 10 &=_2 1010 \\ 15 &=_2 1111 \end{aligned}$$

is an $\mathcal{L}$-position.

(c) $[16, 21, 25]$ or

$$\begin{aligned} 16 &=_2 10000 \\ 21 &=_2 10101 \\ 25 &=_2 11001 \end{aligned}$$

is a $\mathcal{W}$-position and is followed by the $\mathcal{L}$-positions

$$\begin{aligned} 12 &=_2 01100 & 16 &=_2 10000 & 16 &=_2 10000 \\ 21 &=_2 10101 & 9 &=_2 01001 & 21 &=_2 10101 \\ 25 &=_2 11001 & 25 &=_2 11001 & 5 &=_2 00101 \end{aligned}$$

or $[12, 21, 25]$, $[16, 9, 25]$, $[16, 21, 5]$.

2. The characterisation is the same for any number of piles. Write the piles sizes in binary one above the other. The position is $\mathcal{L}$ if and only if each column contains an even number of 1s.

3. Clearly $[0,0,1]$ and $[1,1,1]$ are $\mathcal{L}$-positions, while $[0,0,0]$ and $[0,1,1]$ are $\mathcal{P}$. Now we can exactly mimic the $mex(Z)$ analysis we used for normal nim in problem 5.12. If we draw up tables for normal and misère nim side by side they turn out to be identical apart from the differences already mentioned.

4. It is worth trying to build up our understanding the game using fewer coins.

   With one coin we may represent the state of the game by x_1, the number of empty cells to the right of the coin. If $x_1 = 0$ it is an $\mathcal{L}$-position and if $x_1 > 0$ it is a $\mathcal{W}$-position. This is reminiscent of nim with one pile.

   With two coins we call the number of adjacent empty cells to the right of the first (that is leftmost) coin x_1 and the number of empty cells to the left of the second coin x_2. We claim that the position is $\mathcal{L}$ if $x_1 = 0$. From such a position Alice must always decrease x_2 and increase x_1 and Bob can then respond by reducing x_1 back to zero which ensures that Bob always has a move. If $x_1 > 0$ then the position is $\mathcal{W}$ since Alice can win by reducing x_1 to zero on her first turn.

   With three coins we define x_1, x_2 and x_3 to be the number of adjacent empty cells to the right of first, second and third coins respectively. If $x_3 = 0$ we are essentially playing with two coins and the position is $\mathcal{L}$ exactly when x_1 is also equal to zero. If $x_3 > 0$ and $x_1 = 0$ then the position is clearly $\mathcal{W}$. The next simplest position to consider is one where $x_1 = x_3 = 1$ Alice is sure to lose if she starts by reducing either of these quantities to zero, so her only hope is to move the second coin, thereby decreasing x_2 and increasing x_1. Unfortunately for her, Bob can respond by reducing x_1 back to 1. Eventually x_2 will reach zero at which point Alice is sure to lose. Therefore positions with $x_1 = x_3 = 1$ are all $\mathcal{L}$.

   Now a similar argument shows that positions with $x_1 = x_3 = 2$ are all $\mathcal{L}$-positions. In fact all positions with $x_1 = x_3$ are $\mathcal{L}$. These are the same winning positions as a game of nim played with two piles containing x_1 and x_3 counters respectively. In fact, the only

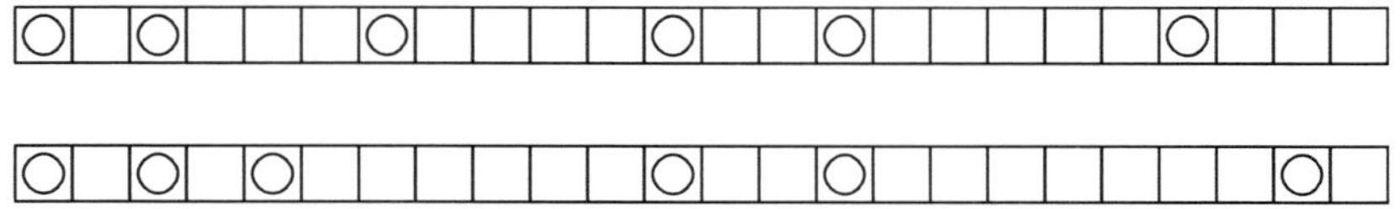

Figure 5d.A

difference between the games is that Alice can increase x_1 a finite number of times, but in each case Bob can respond by reducing it back to its original value.

The analogy is exact. For a game with five coins we define the numbers x_1, x_2, x_3, x_4, x_5 and x_6 in the obvious way, and see that the game is equivalent to the nim position $[x_1, x_3, x_5]$.

In our particular example we have $[x_1, x_3, x_5] = [1, 6, 5]$. The position is $\mathcal{W}$ and in traditional nim there is a unique $\mathcal{L}$-position which follows it, namely $[1, 4, 5]$. However, in this game we may also increase the values of x_1, x_3 and x_5 to a limited extent and we may therefore be able to reach other $\mathcal{L}$-positions. Keeping x_1 and x_3 fixed (and consulting table 5.2 if necessary) we see that we can reach the $\mathcal{L}$-position $[1, 6, 7]$. The $\mathcal{L}$-position with $x_3 = 6, x_5 = 5$ is $\{3, 6, 5\}$ but we note that since $x_2 = 1$ it is not possible to reach this position.

Therefore, there are two $\mathcal{L}$-positions which follow the given position, and these are shown in figure 5d.A.

Exercise 5e

1. It appears that, after some bad behaviour near the start, the sequence settles down into a repeating pattern of length twelve. If this pattern continues, then the next term in the sequence is a 4.

   To prove the pattern does indeed continue, we need to show that $k_{192} = k_{180}$.

   To do this we need to think carefully about how k_n is calculated.

   After one move, the position (n) is split into two different games of Kayles (one of which may be (0), the empty game). If the first game is (r) the second is either $(n - r - 1)$ or $(n - r - 2)$ depending on whether one or two skittles were knocked down. Therefore, to

Start	Followers	Nim-values	a_n
(0)			$mex(\emptyset) = 0$
(1)			$mex(\emptyset) = 0$
(2)	(0)	0	$mex(\{0\}) = 1$
(3)	(1)	0	$mex(\{0\}) = 1$
(4)	(2), (1,1)	1,0	$mex(\{0,1\}) = 2$
(5)	(3), (1,2)	1,1	$mex(\{1\}) = 0$
(6)	(4), (1,3), (2,2)	2,1,0	$mex(\{0,1,2\}) = 3$
(7)	(5), (1,4), (2,3)	0,2,0	$mex(\{0,2\}) = 1$
(8)	(6), (1,5), (2,4), (3,3)	3,0,3,0	$mex(\{0,3\}) = 1$
(9)	(7), (1,6), (2,5), (3,4)	1,3,1,3	$mex(\{1,3\}) = 0$
(10)	(8), (1,7), (2,6), (3,5), (4,4)	1,1,2,1,0	$mex(\{0,1,2\}) = 3$
(11)	(9), (1,8), (2,7), (3,6), (4,5)	0,1,0,2,2	$mex(\{0,1,2\}) = 3$

Table 5e.A

find k_{192} we compute values of the nim positions $[k_r, k_{191-r}]$ and $[k_r, k_{190-r}]$ for $r < 96$ and take the *mex* of all these values.

On the other hand, k_{180} is the *mex* of the values of $[k_r, k_{179-r}]$ and $[k_r, k_{178-r}]$ for $r < 90$.

The question gives us that $k_n = k_{n-12}$ for $83 \leq n \leq 191$.

This means that for small values of r, specifically $r < 90$, we have $[k_r, k_{191-r}] = [k_r, k_{179-r}]$ and $[k_r, k_{190-r}] = [k_r, k_{178-r}]$.

For $90 \leq r < 96$ we have $[k_r, k_{191-r}] = [k_{r-12}, k_{179-(r-12)}]$ and $[k_r, k_{190-r}] = [k_{r-12}, k_{178-(r-12)}]$.

Therefore, the nim-values whose *mex* gives k_{192} are exactly the same as the nim-values whose *mex* gives k_{180}, so $k_{192} = k_{180} = 4$ as required.

The same argument can be used to prove by induction that the repeating pattern continues indefinitely.

2. The required values of a_n are shown in table 5e.A.

3. If we compare table 5e.B and table 5e.A, we see that $b_n = a_{n-1}$ for $1 \leq n \leq 8$. Moreover, the penultimate columns in the tables seem identical.

Start	Followers	Nim-values	b_n
(0)			$mex(\emptyset) = 0$
(1)			$mex(\emptyset) = 0$
(2)			$mex(\emptyset) = 0$
(3)	(1,1)	0	$mex(\{0\}) = 1$
(4)	(1,2)	1	$mex(\{0\}) = 1$
(5)	(1,3), (2,2)	1,0	$mex(\{0,1\}) = 2$
(6)	(1,4), (2,3)	1,1	$mex(\{0,1\}) = 0$
(7)	(1,5), (2,4), (3,3)	2,1,0	$mex(\{0,1,2\}) = 3$
(8)	(1,6), (2,5), (3,4)	0,2,0	$mex(\{0,2\}) = 1$

Table 5e.B

To prove that this pattern continues, we note that b_n is the *mex* of the values of the nim positions $[b_r, b_{n-r-1}]$ for $1 \leq r \leq n-2$. If we assume by induction that $b_k = a_{k-1}$, we find that b_n is the *mex* of the values of $[a_{r-1}, a_{n-r-2}]$ for $1 \leq r \leq n-2$. This is equivalent to $[a_r, a_{(n-1)-r-2}]$ for $0 \leq r \leq n-3$ which is precisely the set of nim positions which determines a_{n-1}.

Therefore $(b_9, b_{10}, b_{11}) = (a_8, a_9, a_{10}) = (1, 0, 3)$.

4. (a) For $n = 1$ white makes the only possible move and wins.

   For $n = 2$ white moves a pawn forward, black captures, white captures and then no more moves are possible so white wins.

   For $n = 3$ white has a choice. She can start by advancing with an edge pawn or the central pawn. In the first case black captures and white will capture back leaving black with a final winning move. In the second case there will be a sequence of four captures, with white making the last move and winning the game. Thus $n = 3$ is a first player win.

   For $n = 4$ white can open with an edge pawn or one of the two centre pawns. In the first case there will be two forced captures just as in the $n = 2$ case. At this point black's position is equivalent to moving first in the $n = 2$ game so he wins. On the other hand. If white opens with a centre pawn, there will then be a sequence of four forced captures. Once these are complete, black has the first move in an $n = 1$ game which he wins. Thus $n = 4$ is a second player win.

Start	Followers	Nim-values	b_n
(0)			$mex(\emptyset) = 0$
(1)	(0)	0	$mex(\{0\}) = 1$
(2)	(0)	0	$mex(\{0\}) = 1$
(3)	(0), (1)	0,1	$mex(\{0,\}) = 2$
(4)	(2), (1)	1,1	$mex(\{1\}) = 0$
(5)	(3), (2), (1,1)	2,1,0	$mex(\{0,1,2\}) = 3$
(6)	(4), (3), (1,2)	0,2,0	$mex(\{0,2\}) = 1$
(7)	(5), (4), (1,3),(2,2)	3,0,3,0	$mex(\{0,3\}) = 1$

Table 5e.C

(b) We will call a column *available* if it still contains two pawns separated by an empty square. A column is *unavailable* if it contains no pawns, or if the pawns it contains cannot move.

The forced capturing rule ensures that if a player advances a pawn in an available column, then that column and any adjacent available columns become unavailable. Forced captures always come in pairs, so unforced moves alternate between the players.

This means that an n-column game of Dawson's chess is in fact equivalent to a variant of Kayles where the initial position is a line of n skittles and a move consists of knocking over a skittle and any of its neighbours that are still standing. This is clearly an impartial game.

(c) We call the nim-value of the position (n) in this game c_n. The first eight values of this sequence are shown in table 5e.C.

(d) It seems that $c_n = a_{n+1}$ suggesting that $(c_8, c_9, c_{10}) = (0,3,3)$.

To prove this pattern continues, we note that c_n is the *mex* of the values of the nim positions $[c_{n-2}]$ and $[c_r, c_{n-3-r}]$ for $0 \le r \le n-3$. By induction these are the positions $[a_{n-1}]$ and $[a_{r+1}, a_{n-2-r}]$ for $0 \le r \le n-3$. Since $a_0 = 0$ we may rewrite $[a_{n-1}]$ as $[a_0, a_{n-1}]$. Now we see that we must take the *mex* set of positions as $[a_r, a_{(n+1)-2-r}]$ for $0 \le r \le n-2$ which is exactly the set of positions which defines a_{n+1}.

5. If the game starts with n crosses, it is not at all obvious that it will ever finish, since the number of free ends is always $4n$. However,

some experimentation suggests that a two cross game always lasts exactly eight moves, making it a win for Bob.

In general we see that no more moves are possible when

(i) The game has only one connected component;

(ii) No region of the plane contains more than one free end.

Initially an n cross game has n components (the crosses) and one (infinite) region. Every turn either increases the number of regions or decreases the number of components. The latter can happen at most $n - 1$ times so the number of regions is always at least $1 + m - (n - 1)$ where m is the number of moves taken so far.

The next key observation is that the rules ensure that every region always contains at least one free end, so the number of regions is always at most $4n$.

Combining these observations shows that $m \leq 5n - 2$.

Moreover, if m is strictly less than $5n - 2$, then either there are two components which can be joined, or there are fewer than $4n$ regions, and therefore a region containing more than one free end. Either way another move is possible, so the game is guaranteed to last exactly $5n - 2$ turns no matter how Alice and Bob choose to play.

Bob wins whenever n is even and loses whenever n is odd.

This result can also be obtained by noting that the completed game is a planar graph with $4n$ faces, $2m$ edges and $n + m$ vertices.

Since we know that in any planar graph $V - E + F = 2$, we obtain $(n + m) - 2m + 4n = 2$, so $m = 5n - 2$ as required.

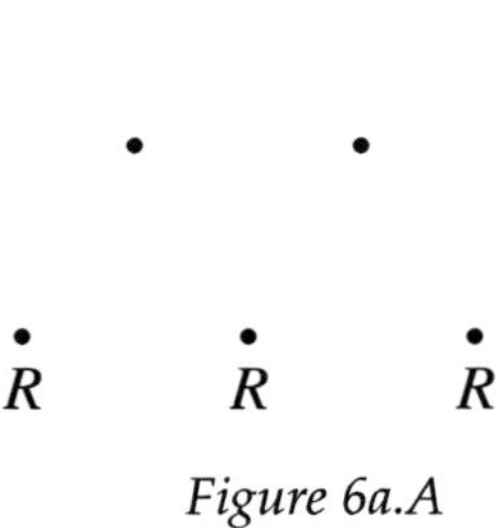

Figure 6a.A

Exercise 6a

1. Consider a grid of points with $k+1$ rows. Each column in this grid contains a colour that is represented more than once. There are k^{k+1} distinct colour patterns for a column, so if the grid contains $k^{k+1}+1$ columns, two columns will have the same colour pattern. These two columns will define a monochromatic rectangle.

2. There are eight ways to colour a column of three points using the colours red and blue. Four patterns are more than half red and four are more than half blue. We claim that four mostly red columns always define a red rectangle. This is because they contain a colour pattern twice, or they contain an all red column and another column with two red points. Similarly, four mostly blue columns always define a blue rectangle. If we have seven columns then we either have four mostly red or four mostly blue columns so we are done.

   For the second part we simply colour the six columns of the grid with the six non-monochromatic patterns and check that there is no monochromatic rectangle.

3. Start with a set of three evenly spaced points which are the same colour and assume, without loss of generality, that they are red. Now consider the set of six points shown in figure 6a.A. If any of the unlabelled points are red, then there is a red equilateral triangle, and if none of them are, they form a blue equilateral triangle.

4. The height of an equilateral triangle with side length one is $\frac{\sqrt{3}}{2}$. We colour the plane with red and blue vertical stripes of width $\frac{\sqrt{3}}{2}$ taking care that the lines either side of a strip are always different

colours. More specifically we colour (x, y) red if there is an integer k such that $\sqrt{3}k \leq x < \sqrt{3}k + \frac{\sqrt{3}}{2}$. Every equilateral triangle with side length one unit has a side which makes an angle of at most 30° with the x-axis. It is now easy to check that the ends of such a segment are always different colours in our colouring.

5. We consider a grid of points with nineteen rows. Each column contains ten points that are the same colour. There are 2^{19} ways to colour a column of nineteen points with two colours. So, if we have $9 \times 2^{19} + 1$ columns in our grid, then ten columns will have identical colour patterns. These columns form the lines parallel to the y-axis, and we can then choose ten lines parallel to the x-axis to ensure that all the intersections are the same colour.

6. Construct a graph where the seventeen correspondents are the vertices, and colour the edge joining two correspondents either red, blue or green, depending on the topic they discuss in their letters. Just as in problem 6.4 we wish to prove that a monochromatic triangle always exists.

   Choose any vertex v. It is contained in sixteen edges, so we can find six that are the same colour. Assume, without loss of generality, that the edges $\{v, v_1\}$, $\{v, v_2\}$,...,$\{v, v_6\}$, are all red and consider the vertices $\{v_1, v_2, \ldots, v_6\}$. If $\{v_i, v_j\}$ is red, then v, v_i and v_j form a red triangle. Otherwise, we have six vertices which are joined by edges of only two colours. However, since $R(3,3) \leq 6$, these six vertices contain a monochromatic triangle as required.

   It turns out that the bound is exact: the number 17 in the problem cannot be reduced. To prove this we would need to exhibit a colouring of the edges of K_{16} which used three colours and contained no monochromatic triangle. We will not do so here, but enthusiastic readers are welcome to try their hand at the task. Those who are curious, but have other demands on their time, might like to look up the Clebsch graph.

Exercise 6b

1. We suppose the vertices of the K_8 we must colour form a regular octagon of side length one and measure distance around the outside of the octagon. Thus the distance between distinct vertices is 1, 2,

3 or 4. We colour edges of the K_8 red if the distance between their end points is 1 or 4. It is easy to check that there are no red triangles. If the distances between vertices of a quadrilateral are a, b, c and d, then we either have $a + b + c = d$ or $a + b + c + d = 8$. The first equation cannot be solved if all the variables are 2 or 3. The only solution to the second equation using 2 and 3 is $2 + 2 + 2 + 2 = 8$, and this square has red diagonals.

2. (a) Three. The argument used in problem 6.5 shows that if a vertex has four red-neighbours or six blue-neighbours, then there will be a red K_3 or blue K_4.

   (b) There must be 9×3 ends of red edges in the graph. However this is impossible since each red edge has two ends and 27 is an odd number.

   (c) This proves $R(3,4) \leq 9$ and hence that $R(4,4) \leq 18$.

3. Using problem 6.9 on page 206 gives $R(s,s) \leq \binom{2s-2}{s-1}$. We prove that $\binom{2s-2}{s-1} \leq 4^{s-1}$ by induction on s.

   The base case ($s = 2$) is easy to check. Now we consider

$$\binom{2s}{s} = \frac{(2s)!}{(s!)^2} = \frac{2s(2s-1)}{s^2}\binom{2s-2}{s-1} < 4\binom{2s-2}{s-1} \leq 4 \times 4^{s-1}.$$

   The last inequality uses the induction hypothesis and completes the inductive step.

4. We seek a colouring of K_{13} with no red triangle of blue K_5. If the vertices form a regular 13-gon then the distance (round the outside) between distinct vertices is an integer between one and six. We call the distance between the end points of an edge its *length* and colour edges of length one or five red. It is easy to check there is no red triangle. If a, b, c, d and e are the lengths of the sides of a pentagon then $a + b + c + d = e$ or $a + b + c + d + e = 13$. If all the sides of the pentagon are blue, then all the lengths belong to the set $\{2, 3, 4, 6\}$. The equation $a + b + c + d = e$ clearly has no solutions from this set. The equation $a + b + c + d + e = 13$ has two types of solution: we might have two 2s and three 3s or three 2s, a 3 and a 4. In either case the pentagon will have adjacent sides of length two and three, so will have a diagonal of length five.

5. Suppose the edges of K_{19} are coloured red or blue. Since $R(2,6) = 6$, if any vertex has six red-neighbours, then the graph has a red K_3 or a blue K_6. Since $R(3,5) \leq R(3,4) + R(2,5) = 14$, if any vertex has fourteen blue-neighbours, then the graph has a red K_3 or a blue K_6. Therefore, if the graph does not contain a red K_3 or blue K_6 then every vertex has exactly five red-neighbours and thirteen blue-neighbours. This means the number of red edge ends is 5×19. This is an odd number which is impossible since every edge has two ends. Hence every 2-coloured K_{19} has a red K_3 or blue K_6

6. We divide $(s-1)^2$ points into $(s-1)$ classes each containing $(s-1)$ points. We join two points by a red edge if and only if they are in the same class. This is a colouring of $K_{(s-1)^2}$ and any set of s points contains both two points in the same class and two points in different classes, so it is not monochromatic.

7. We seek a colouring of K_{17} with no monochromatic K_4. If the vertices form a regular 17-gon then, using the same definitions as question 4, the length of an edge is an integer between one and eight. We colour edges red if their lengths are one, two, four or eight.

   If the lengths of the sides of a quadrilateral are a, b, c and d then either $a+b+c = d$ or $a+b+c+d = 17$. The red solutions to these equations are $1+1+2 = 4$, $2+2+4 = 8$ and $1+4+4+8 = 17$ while the only blue solution is $3+3+5+6 = 17$.

   We now check that if a quadrilateral has red sides, then it has a blue diagonal and vice versa.

   Quadrilaterals with red perimeters have the following side lengths: $1,1,2,4$ or $2,2,4,8$ or $1,4,4,8$. The first type have diagonals of length three, the second have diagonals of length six, and the third have diagonals of length five.

   Quadrilaterals with blue perimeters have sides with lengths $3,3,5,6$. These quadrilaterals all have diagonals of length eight.

Exercise 6c

1. We can mimic the solution to problem 6.13 to show that $R(s,t,u)$ is less than or equal to $R(s, R(t,u))$. Since we have already shown that the latter number exists, $R(s,t,u)$ is certainly finite.

2. We have that the following chain of inequalities.

$$R(s,s,s) \leq R(s, R(s,s)) \leq R(s, 4^{s-1}) \leq \binom{4^{s-1}+s-2}{s-1}.$$

Now we use the facts that $4^{s-1}+s-2 < 4^s$ and that $\binom{n}{r} < n^r$ to obtain $R(s,s,s) < (4^s)^{s-1}$ which is stronger than the required inequality.

3. It is easy to adapt the argument used in problem 6.11 to show that if $\binom{n}{s}3^{1-\frac{s(s-1)}{2}} < 1$, then $R(s,s,s) > n$. To finish the question we must show that if $n \leq \sqrt{3}^s$, then

$$\frac{n(n-1)\dots(n-s+1)}{s!} < 3^{\frac{s^2}{2}} \times 2^{-\frac{s}{2}} \times \frac{1}{3}.$$

As in problem 6.12, it now suffices to show that

$$n^s < 3^{\frac{s^2}{2}} \times (s! \times 3^{-\frac{s}{2}} \times \frac{1}{3}).$$

We can check (by induction or otherwise) that the expression in parentheses on the right hand side is greater than one for all S. The condition that $n \leq 3^{\frac{s}{2}}$ implies that $n^s < 3^{\frac{s^2}{2}}$ so we are done.

4. We have that $R(s,t,u,v) \leq R(R(s,t), R(u,v))$, and we have already proved that the number on the right hand side is finite.

5. We have the following chain of inequalities.

$$R(s,s,s,s) \leq R(R(s,s), R(s,s)) \leq R(4^{s-1}, 4^{s-1}) \leq 4^{(4^{s-1}-1)} < 4^{4^s}.$$

Exercise 6d

1. We call a colouring of $[N]^{(2)}$ *good* if it contains a nice set and *bad* otherwise. We need to prove that, for all sufficiently large N, every colouring is good.

   We suppose, for contradiction, that for each i there is a bad colouring of $[i]^{(2)}$ which we call C_i^1.

   We now consider the sequence of colourings $C_1^1, C_2^1, C_3^1, \dots$ and use them to construct a colouring C on $\mathbb{N}^{(2)}$ just as we did in section 6.7.

This colouring contains an infinite monochromatic set with least element v_1. Suppose the first v_1 members of this infinite set are all members of $[n]$ for some n (we may take $n = v_{v_1}$). Suppose further that every edge in $[n]^{(2)}$ is among the first M edges in the list of edges used to construct C. This means that the colouring C_1^M coincides with the colouring C on $[n]$, which, in turn makes C_1^M a good colouring. Since all the colourings in the list were bad, this is a contradiction.

Bibliography

[1] Paul Erdős and George Szekeres. "A combinatorial problem in geometry". In: *Compositio Mathematica* 2 (1935), pp. 463–470.

[2] Paul R. Halmos and Herbert E. Vaughan. "The Marriage Problem". In: *American Journal of Mathematics* 72 (Jan. 1950), pp. 241–215.

[3] Lefteris Kirousis and Georgios Kontogeorgiou. "The *problème des ménages* revisited". In: *The Mathematical Gazette* 102 (Mar. 2018), pp. 147–149.

[4] Gerry Leversha and Dominic Rowland. *Introduction to Combinatorics*. Handbooks Six. UKMT, 2015. ISBN: 978-1-906001-24-7.

[5] Jeff Paris and Leo Harrington. "A Mathematical Incompleteness in Peano Arithmetic". In: *Handbook of Mathematical Logic*. Ed. by Jon Barwise. North Holland, 1970.

[6] Frank P. Ramsey. "On a problem in of formal logic". In: *Proceedings of the London Mathematical Society* s2-30 (1930), pp. 264–286.

[7] Elwyn R.Berlekamp, John H. Conway, and Richard K. Guy. *Winning Ways for your Mathematical Plays (Volume 1)*. A.K. Peters, 2001. ISBN: 1-56881-130-6.

Index